HUMAN CONSCIOUSNESS
THE IMPACT OF LANGUAGE AND CULTURE

RAFAEL PINTOS-LÓPEZ

For Inés who, as usual, had to put up with the lunacy and the obsession. She helped and edited and revised.

For Hernán, Rodrigo and Millán, whom I admire. They happen to be my sons.

CONTENTS

Preface vii
Introduction xv

1. Currently in Neuroscience 1
2. A Philosophy of Consciousness 16
3. Language, Culture and Consciousness 37
4. Intersection with Biology 58
5. Communication 64
6. Time and Culture 70
7. Translation 86
8. Need for a Definition 92

Conclusion 97
Acknowledgments 107
Some Quotations 109

PREFACE

*"On a withered branch,
a crow has come to perch—
at dusk in autumn."*

You have just read a poem written long ago. The author's name was Matsuo Kinsaku, but he was mostly known as Bashō. The poem is signed Tōsei, another name he was known by, a pseudonym. But that is irrelevant. He wanted you and me to visualise a black bird—a species that you and I know—a crow, on a branch at sunset.

You and I, the readers, can imagine the beauty of the bird and the dusk that Bashō saw that autumn. You and I know what a poem is. This type of poem is called a *haiku*.

The symbols he wrote weren't letters. They were a mixture of ideograms (*kanji*) and syllables (*hiragana*).

They looked more or less like this:

PREFACE

枯枝に
からすとまりたるや
秋の暮桃青

A moment ago, in your mind, you had sounds. Those were English language sounds. The poem sounded more like this:

> *"Kare eda ni*
> *karasu tomaritaru ya*
> *aki no kure."*

The poem is deceptively simple. Bashō was trying to describe—as directly as possible—nature and the beauty of what happened that sunset. The crow wasn't special, it had no name. It was *the* crow, the essence of one, a symbol of its species. And the branch was old and dry; it could have been anywhere. You and I have seen many like it. Bashō described a moment in nature.

The *haiku* was translated into English by John T. Carpenter.

You could be reading this in Norway, maybe this year; maybe in South Africa, sometime in the future. But you can visualise the bird and the branch just like me, and something that happened at sunset one autumn, long ago, in Japan. Something that Bashō saw and felt, centuries ago, and that he wanted you to see and feel within your mind. The beauty of nature as perceived by a fellow human being.

You understood it through John T. Carpenter, who translated it I don't know when, and through me, who transcribed it here. I am writing this in Queensland, Australia, in 2024.

This is how we humans communicate. We can do it beyond, time and space, language and culture. The memory of that moment is nowhere, except in those symbols.

But how do you call that process? How do you call a process whereby we can communicate information, transmit ideas and induce feelings among humans, a process that unites us in our humanity?

It is not physical. It is an intangible quality that we possess. Something we acquired aeons ago, when we acquired language and became meta-evolutionary.

The process is human consciousness. We all possess it and we are all part of it.

It cannot be understood by analysing processes within our physical brains.

What you are about to read—this book—is all about human consciousness.

There are things one has to do. This is the offshoot of two previous books I wrote about human consciousness and it includes some of their sections; the books are: *Consciousness and Time - A New Approach* (2023), and *On the Emergence of Human Consciousness* (2023). There is no pretence of scientific research here, nor is it an academic book. This is a lay book, a book for a public interested in the mysteries of human consciousness; like the previous ones, it is the fruit of a heuristic process, a commonsense reinvention of the wheel, if you want to call it that.

PREFACE

The title of the book sums it all up. It's a direct answer to the current physicalist approach of the scientific world. Neuroscientists disregard the idea of a consciousness unique to humans, one that would include language and culture. Language and culture, however, are an important part of human consciousness. Without those two elements, human consciousness is animal consciousness that has evolved. But there is no real continuity. At the onset of humanity, we acquired language and became meta-evolutionary. This book questions how science can account for the metaphysical component of consciousness when focusing solely on individual brains.

Science wants the intangible part of collective human consciousness to be ignored "until further notice". The problem is that we are not a collection of individuals. There is much more to humanity. There is a clear '*gestalt*', a special synergy, that applies to the growth of humanity and that cannot be reached or explained by studying individual brains, and that is what neuroscience is doing (and making—I must admit—incredible progress in other ways). Neuroscience has been actually studying the brain, pretending that a physical organ can generate human consciousness.

But going back is impossible for humanity. After the introduction of language and the enlargement of human groups—which eventually became civilisations—the intelligence of *H. sapiens* grew exponentially. A line graph with time and human intellectual/demographic growth variables would show a horizontal line for hundreds of thousands of years and an almost vertical line since the introduction of language.

I just needed to write this third and final book because in it I emphasise, even further, that language and culture are important phenomena that largely influence human consciousness. Perhaps it is a reaction, as I state above, to a scientific world that cannot find its bearings.

The way I conceive human consciousness and its emergence places my thought within Cartesian dualism. I never intended any such thing. The position adopted is based on what I believe is evident and—as stated above—commonsense. Things are the way they are. Descartes wouldn't have been Descartes without cognition and metacognition. He knew he was somebody because he was—first and foremost—a _human_ being; he was part of this species and part of its collective consciousness.

The fact is that, currently, very few neuroscientists, philosophers, or researchers in the field of consciousness accept a dualist position as something remotely tenable, or even deign to consider that consciousness may have a hybrid nature. In the current study of consciousness, that is akin to stating that a flat planet Earth lies on a pillar of giant turtles that goes "all the way down". Unfortunately, what I state here is far from "scientifically fashionable". However, I know that—eventually—the pendulum will have to swing the other way. In fact, it already is swinging away from reductionism. Frank, Gleiser and Thompson (*The Blind Spot*) are at the vanguard of that movement of scientific renovation:

"The Universe and the scientist who seeks to know it become lifeless abstractions. Triumphalist science is actually humanless, even if it springs from our human experience of the world... Scientific knowledge isn't a window onto a disembodied, God's-eye perspective. It doesn't grant us access to a perfectly knowable, timeless objective reality, a 'view from nowhere', in philosopher Thomas Nagel's well known phrase. Instead, all science is always our

science, profoundly and irreducibly human, an expression of how we experience and interact with the world.".

There are other authors like them who are brave enough to question reductionism and objective reality. It's a beginning.

Science has forever been in a physicalist and reductionist mode. Cartesian dualism—i.e., stating that mind and matter are separate entities—is not acceptable, even if you explain that they are inseparably integrated.

Let's try to clarify what we have seen so far.

Science has advanced a lot; nobody can deny that. Being in favour of science does not mean, however, believing in 'scientism'—the current dogma—which basically states that the only way to establish what is true or not true is through the scientific method, that the truth has to be tested by means of experiments. That applies especially to the study of consciousness. But that study, in particular, shows that science has reached its uppermost limit. After decades of trying to prove its hypothesis, it has failed to find the correlates of consciousness in the brain. That would mean that probably you cannot study consciousness from a scientific perspective alone.

The lay person doesn't understand that and, unfortunately, has been misguided to this blind belief in science. After Dawkins sold the general public his own version of scientism, science has replaced religion. Unfortunately for the believers, there is a metaphysical aspect of reality that needs to be studied as well. That is where philosophy comes into the picture.

PREFACE

Some philosophers believe—with neuroscientists—that consciousness emerges from neurones in the brain. There are some who believe that consciousness comes first and reality emerges from it. Some of the latter ones even subscribe to the theory of panpsychism, which states that consciousness is everywhere in the universe, that everything is conscious, including inanimate things. Those assertions cannot be falsified.

There is another school, not very popular nowadays—mostly rejected as absurd—that believes neither. They are the Cartesian dualists. They believe that body and mind have different natures. Dualism is not reductionist. Reality, it affirms, is neither physical nor metaphysical, but both at the same time.

So, scientists do not believe in anything metaphysical because they cannot study it. To them, it is "mumbo jumbo". Philosophers are against dualism because it does not reduce reality to just one nature. Some philosophers who are against it do so on the basis of its lack of elegance. Two natures of reality? That has to be wrong, they say. Things have to be reduced to their minimum exponent.

I say that there are fundamental aspects of nature that you cannot reduce. Elegance does not necessarily mean reductionism. Etymologically, 'elegance' derives from Latin 'eligere', to select carefully, to choose. Sometimes, when things are as intertwined as cognition and sentience within human consciousness, you cannot choose anymore.

Objective reality has limits, and human consciousness is one of them. The current school of 'orthodox' philosophy is reductionist, but reality cannot afford any further reduction: body and mind are entangled and layered—if you like—and they do not share the same nature. One is physical and the other one, metaphysical. The human mind is where subjective and objective meet.

Here I explain why; the reader can decide whether what I say has any merit, whether some hypotheses are right and some are wrong.

R P-L.

INTRODUCTION

"The great cognitive shift is an expansion of consciousness from the perspectival form contained in the lives of particular creatures to an objective, world-encompassing form that exists both individually and intersubjectively. It was originally a biological evolutionary process, and in our species it has become a collective cultural process as well. Each of our lives is a part of the lengthy process of the universe gradually waking up and becoming aware of itself".

- Thomas Nagel

If we are going to discuss human consciousness, let's begin with some facts that cannot be ignored: human beings speak and live within cultures. Human beings are, to a large degree, products of their culture.

Some animals—mostly mammalians—can communicate (not in any way comparable to human language) and share some cultural traits within their groups. Chimpanzees are a clear example; elephants, who behave following cultural rules, are also among animals that communicate to a certain extent, and appear to have a fairly basic culture. Of course, there are many other species that behave according to collective parameters.

If we believe that human consciousness is just an evolutionary development of our biological nature; if we believe that the minds of both, animals and humans, share the selfsame nature, then we also believe that the current physical 'neuroscience' approach to the study of human consciousness is the correct one and will bear fruit soon.

If we believe, however, with Thomas Nagel, that there has been a shift in the nature of consciousness and that, in humanity *"... [what] was originally a biological evolutionary process, ... in our species it has become a collective cultural process as well."*, then, we must also believe that we need a change of paradigm. The study of human consciousness cannot be restricted to the physical approach that neuroscience has been using for decades with no real result in terms of finding the 'correlates of consciousness in the brain'. What happens within the individual brain is definitely not going to give us an answer. We have now an *"objective, world-encompassing [reality] that exists both individually and intersubjectively"*. It will be impossible to reach that intangible reality concentrating solely on the individual brain.

Henri Bergson, who had a complicated relationship with evolutionary thought, began *Creative Evolution* as follows:

"The history of the evolution of life, incomplete as it yet is, already reveals to us how <u>the intellect has been formed, by an uninterrupted progress</u> (my underlining), along a line which ascends through the vertebrate series up to man. It shows us in the faculty of understanding an appendage of the faculty of acting, a more and more precise, more and more complex and supple adaptation

of the consciousness of living beings to the conditions of existence that are made for them.".

During the twentieth century many—like him—accepted the evolutionary nature of human consciousness as part of the prevailing dogma.

Then they found the *'hard problem'*. Philosopher David Chalmers sometimes talks about a *'complex consciousness'*—i.e., the human one—and a *'simple consciousness'*—that is, animal consciousness. In doing so, he suggests that both share the same nature, when they do not. Perhaps, to avoid that sort of confusion, it would be preferable to allocate different names to them, as they differ radically. Animals are sentient beings. We are rational, cognitive beings, the nature of our consciousness includes sentience, but includes cognition as well; it is hybrid. It has an extra layer, which is intangible and collective. Why not call 'simple consciousness' 'sentience' and differentiate it from 'complex [human] consciousness' on that basis? Why not take sentience as 'fundamental' in nature? Sentience is what is needed for life. All living creatures need sentience. If we're interested in human consciousness, perhaps we should leave it at that for the time being.

When we discuss whether LLMs (Large Language Models) or AI (Artificial Intelligence) will become conscious, we are currently deluding ourselves. LLMs will probably end up really *understanding* (which at present they do not)— which is a huge step in the development of intelligence—, but they will never become conscious in their current state. They're just a human creation. They lack sentience. To be really conscious, the way a human being is, apart from intelligence and understanding, you need the first basic step: sentience. You need it to be alive.

INTRODUCTION

We have created fictional characters, like the Golem, or Frankenstein's monster, who did not meet the requirements to be conscious. They lacked what the ancient Hebrews called *'nephesh'*, the breath of life. Without it, sentience does not occur. Maybe at one point in the distant future, we will be able to create hybrid, intelligent beings, like *Bladerunner's 'replicants'*. AI is far from that stage of sophistication.

But, going back to human consciousness, we have disciplines called 'Humanities'. The name says it all: they study human beings. The nature of human consciousness is hybrid: physical *and* metaphysical. The shift in research will have to be from the purely scientific approach to a hybrid one, but conducted from an all-encompassing perspective. The beginning of humanity is marked by an exponential growth in cognition; what's more, our consciousness is still growing. Our collective knowledge is immense; it cannot be compared to that of any individual. The centres for the study of consciousness need to be basically humanistic with some scientific input, not the other way around.

After the big leap into culture, our species—*H. sapiens*—began to communicate and think. We are conscious and understand a lot about the universe, much more than any other species. The human metacognitive process originated in language and culture; that is easily verifiable.

The more you study human consciousness, and the more you read about it, the more you will come to one conclusion: we are an amazing species. We have been extraordinarily successful. We are unique. These are not exaggerations. The fact that there are almost eight billion human individuals conclusively proves it. No other mammal has reached anywhere near those figures. *H.*

sapiens has taken over the planet. We discover, we create, we invent.

There is much more to us than to any other animal species. Why do I say this? Well, we have gone way beyond biological evolution, where no other animal group has gone. We are the only meta-evolutionary species. We have developed recursive, complex languages; we have developed intricate cultures with countless rules, and we enjoy an individual consciousness which is integrated into the collective consciousness I mention above, that allows the reader to decipher the symbols I am typing into this laptop. We can communicate the most complicated of notions. We can co-operate in all kinds of magnificent enterprises. Our cognitive abilities are unparalleled. Arguably—but most probably—we are the sole witnesses of the universe.

So far I have used terms that may appear exaggerated to the sceptic: 'meta-evolutionary', 'unique' and 'unparalleled'. Whether other species—like *H. neardenthalensis*—may or may not have had the basics of a communication system is immaterial; the exponential growth in intelligence that *H. sapiens* experienced ever since a human individual understood a sound produced by another human deserves those adjectives.

I believe the *H. sapiens* species already underwent a technological singularity. I believe our species became a 'black swan' phenomenon. We are now riding the wave of that exponential growth (which some say will accelerate even more, into a spiral of change).

The instant human beings encountered language is significant from any possible perspective. The impact that moment had on the cognitive component of human consciousness is something that deserves the special attention it has not received thus far.

INTRODUCTION

When I use the term 'meta-evolutionary', I do it with certainty, convinced that it is the correct expression. The tipping point in the evolution of our species occurred when the 'fittest individual' of the species ceased to be the strongest one. The challenge to the biological fitness of the species came about when random mutations, or other forms of gene change, ceased to be the element that increased the rate of growth of the species. The focus of growth of the species became, then, the individual that could transmit information to other individuals most effectively in order to convince them to co-operate with them—in projects or otherwise. At that point, language, culture, and technological advancement became the forces that propelled *H. sapiens* towards exponential growth.

How did it happen? How is it happening? What I propose is a commonsense dualist solution: I argue that human consciousness includes two components, or layers: a basic one and a high one, and that they are discrete, evolved from different needs of our species, that arose during different periods. I prefer to think of them as layers, as the term implies a more visual way of understanding that the high one is superimposed on the basic one. Literally—from a physical point of view—because the neocortex is a massive neural addition that envelops the less developed areas of our triune brain (as Carl Sagan would have called it), or, if you prefer, our adaptive brain. The hypothesis proposed here fits both descriptions of the brain. What is proposed is that, in humans, sentience and cognition are discrete but interdependent.

As I stated previously, humans might not be the only species to have discrete layers (in terms of sentience and some cognition).

A recent study on macaques conducted by an international team headed by Sean Froudist-Walsh at the University of Bristol, confirmed that there is a separation between sensory and cognitive networks in other species as well. The fact remains that human consciousness has exponentially outgrown those of all other species.

One of the most evident qualities of the individual brain is that it processes thought. The human brain—like that of other animals—is a physical entity, an organ of the body. That has led to a concentration of the study of consciousness within the biological sciences. Neuroscience focuses its research within the physical limits of the brain, from a biological perspective. The problem with that approach is that the most important component of human consciousness, cognition, is not biological, but cultural. What's more, it is not physical but intangible.

Brain receptors appear to operate on the basis of patterns that shed light on their function. This happens in terms of perception, memory, and emotion. The layers, however, cannot be peeled off to be analysed separately. They are distinct, but integrated to a large extent.

In any case, studying one type of layer will not help solve the problems associated with the other, as they are two different phenomena, discrete and overlapping, not a continuum. Of course, the logical conclusion in that respect is that human consciousness cannot be explained by reducing it to its physical constituents. Feelings, sensations, qualia, are essential, indispensable survival tools. All living beings must have them. That should be considered another problem, another field of study. Qualia should be considered, perhaps, a fundamental quality of being alive.

INTRODUCTION

∼

In their brilliant book,*The Blind Spot: Why Science Cannot Ignore Human Experience* (2024), which I quote extensively here, Adam Frank, Marcelo Gleiser and Evan Thompson suggest there is a strange loop:

"We're now confronted with a strange loop. Horizonal consciousness subsumes the world, including our body experienced from within, while embodiment subsumes consciousness, including awareness in its immediate intimacy. The primacy of consciousness and the primacy of embodiment enfold each other. We need to examine this strange loop which disappears from view in the Blind Spot.".

Their hypothesis takes them later to avoid the loop by suggesting that what they call life-world—*breath of life*, or what the ancient Hebrews called *nephesh*—has primacy over consciousness and embodiment, which is, indeed, irrefutable.

For the purposes of the study of consciousness, I treat life-world/*nephesh* as fundamental. But I reject the idea of a consciousness-embodiment loop, as I view those phenomena chronologically. Life-world precedes embodiment, which subsumes sentience (an integral part of consciousness), and to it humans added cognition, a result of the growth of language and culture. My metaphor of a layered human consciousness should not be taken literally but, I believe, helps understand the fact that cognition is a later addition to sentience. Those two components, discrete, but intertwined and operationally together, produce what we now understand as human consciousness.

At another point, Frank, Gleiser & Thompson touch briefly on whether life precedes basic animal consciousness or not—

which, if you exclude panpsychism from the equation, does not appear to be viable at all:

"We need to explain how the emergence of conscious organisms happened, including whether sentience is coeval with life or a later evolutionary event, and whether the emergence of consciousness is accidental or inevitable from an evolutionary standpoint.".

And then they go on to provide an explanation themselves:

"Consciousness is subjective and experiential, whereas biological life is construed as entirely objective".

The chronological sequence of the two is quite evident: [except for the pansychist hypothesis] you cannot have consciousness without life. I do understand the context in which they pose those questions. Somehow, thinking about conscious amoebas at this stage is not going to help us solve the problem of human consciousness. Yes, we do need to do what they suggest. But I believe treating life as fundamental at this stage would allow us to concentrate on the real study of the human mind. All evidence is that sentience (and perhaps a germ of cognition in some mammals) preceded human consciousness, and that the latter really appeared with the meta-evolutionary introduction of language.

The acquisition of human consciousness took some aleatory turns. Our species arrived at it through our relative physical weakness and our need for a lengthy upbringing. Like some birds, we are an altricial species: we are born extremely immature and require maternal and collective care for many years.

INTRODUCTION

In our current societies, infants need to be taught language, information, literacy and numeracy, and they have to be socialised, before they can become full members of the collective, or culture. These abilities appear to have increasingly evolved from the moment humans began living in larger groups, thanks to language, trust, and food-sharing, and have indeed transcended standard evolutionary biology.

But I used the term 'altricial' in this Introduction, and now I intend to digress a few paragraphs, analyse, and explain in more detail what I meant by that. Altricial animals are unable to care for themselves for a long time. Some bird species are altricial; kangaroos, dogs and cats also need prolonged care. Like those species, we are born helpless, and our infancy, childhood and teenage years are lengthier than those in any other species. But we go beyond the biological.

In her article 'How to grow a human', Brenna Hassett, from the University of Central Lancashire, ponders the reasons for our lengthy upbringing.

"The average human spends at least one quarter of their life growing up. In the careful calculus of the animal kingdom, this is patently ridiculous. Even most whales, the longest of the long-lived mammals, spend a mere 10 per cent or so of their time growing into leviathans. ... Could the secret to our species' success be our slowness in growing up? And if so, what possible <u>evolutionary benefit</u> (my underlining) *could there be to delaying adulthood – and what does it mean for where our species is going?"*.

Hassett questions the classic explanation for our success—which has long been thought of as our ability to walk upright and our big brain—and proposes a new alternative, to but she also does it from an evolutionary perspective. We spend a lot of time teaching our children how to survive and thrive in society, and that includes language and all sorts of cultural information. She continues by comparing our growth, from the 25-odd years it takes for us to reach physical maturity, to that of our nearest relatives and concludes that we are a decade or so slower.

Hassett also compares the growth of current children to that of two ancestors: the Taung child and the Nariokotome boy. Her conclusion is that *"Our long childhood is a uniquely evolved human trait"*. And she is correct. She is, again, correct when she discusses a unique human institution: the grandparents (and, I would add, the father and the extended family) in terms of the help required to raise children. Then, she asks:

"All the unique qualities of human childhood are marked by this kind of intense investment. But that raises the big question. If 'winning' evolution looks like successful reproduction, then why would we keep our offspring in an expensive holding pattern for longer than necessary?".

Of course, human children require massive amounts of information that other mammalians don't need because human children have to live within cultures that have highly sophisticated regulations and language.

The development of this new type of consciousness was refined and exponentially accentuated as groups of humans became larger and larger. And it continues to develop and become more sophisticated every generation. The more successful and the more gregarious we became the more human consciousness grew. In the process, it acquired metacognition, elaborate thought, long-term memory, creativity, individual and social identity, and time, among other qualities and concepts.

INTRODUCTION

We can state, perhaps, that taking a long time to grow up does not mean then that our species wastes time. The process is absolutely necessary.

Going back to the beginning of this introduction and trying to summarise it in dot points, the hypothesis is as follows:

* Human consciousness consists of two integrated but discrete layers:

1) basic animal consciousness, or sentience.

2) high human consciousness, or psyche.

* We are born sentient. (For expedience sake, this book treats sentience as a fundamental component of consciousness). High consciousness is only acquired through parental and collective upbringing. It is collectively and individually transmitted. Its nature is cultural.

* To some extent, this appears to confirm the Sapir-Whorf hypothesis of linguistic relativity, and falsifies universalist claims concerning language, as proposed by Noam Chomsky.

* Understanding the workings of high consciousness cannot be arrived at through an evolutionist study of sentience, as it will always reach the biological ceiling.

* Imagination, creativity, language, adventurousness, are exclusively human traits acquired through high consciousness, i.e., by cultural means.

* The logical conclusion is that there should be cortical and other centres in the brain, newer than any centre that deals with

strictly biological phenomena, where cultural developments have been integrated and are processed (e.g., Broca's and Wernicke's areas).

* Time is a human construct that exists only within high consciousness, through unlimited imagination (expectation) and long-term memory (which involves identity and collective perception). Change is universal. Time is cultural.

* Without high consciousness there is only present and change.

Let us see how we can clarify all of these hypotheses.

Of course, the ideas as explained here are very basic; they would need to be explored and analysed more thoroughly, in a more exhaustive fashion. The implication is that the main requirement would be to separate seed from chaff without being reductive. A change of paradigm along these lines may help the study of consciousness.

CURRENTLY IN NEUROSCIENCE

"Finally, any attempt to deny that your immediate awareness has outside sources is incoherent, because this denial cannot be thought or stated except through the conceptual and linguistic resources provided to you by your social heritage and your cultural and biological evolution. Even if you were the last surviving person of a universal plague, your mind and your capacity for meta-awareness would be inherently social."

- Adam Frank, Marcelo Gleiser & Evan Thompson

What the Introduction proposes is plain commonsense—or so I believe; the scientific world, however, is unlikely to accept any dualistic proposal at this stage. Neuroscience keeps on studying individual brains in the hope that it will find the place where consciousness emerges. There is no such place. Our brains have a neocortex, which is the product of our cultural consciousness through a feedback loop. Our neocortex appears to have grown from the moment we acquired our distinct human consciousness, i.e., its origin, and spectacular growth, seems to be cultural. In any case, even if neuroscientists find a physical place, how will the study explain the 'emergence of consciousness'?

Individual human consciousness is not purely physical. It does not emerge from the brain or in the brain; it's the other way around: neurones are created as a result of behaviour.

Apart from the neocortex, the human hippocampus has grown, no doubt, because we have more cognitive functions related to memory, whereas the memory areas to do with emotion have decreased. That is noticeable when compared with those of other primates.

A 2020 article by Rogers Flattery *et al* gives an account of the proportional growth of the learning and long-term-memory-related hippocampus area in humans as opposed to emotional memory:

"The hippocampal formation is important for higher brain functions such as spatial navigation and the consolidation of memory, and it contributes to abilities thought to be uniquely human... In the context of prior investigations of rhesus monkeys and humans, our findings indicate that, in the hippocampal formation as a whole, the proportions of neurones in CA1 and the subiculum progressively increase, and the proportions of dentate granule cells decreases, from rhesus monkeys to chimpanzees to humans."

What happens is that the neocortex and the hippocampus are part of an information system that individuals share with the rest of humanity. Collective consciousness was the effect of culture and—amazingly—we grow from infancy to become integrated into that common culture.

In another recent article, Ben Turner describes the findings of a study conducted by physicists at the University of Sydney. The study discovered certain mysterious wave patterns in the neocortex. Turner reports:

"The wrinkled, outermost layer of the brain —known as the cerebral cortex— manages many of the mind's most complex tasks, such as memory, attention, language, perception and even consciousness itself ... Yet neuroscience has mainly ignored the cortex itself and instead traditionally focused on the connections and interactions between neurones (the brain's nerve cells) to determine how the wrinkly organ functions."

What he is saying is that neuroscientists appear to ignore the outer cortex—where most of human consciousness resides—and concentrate on solely biological functions. Of course, this physicalist bias—that concentrates not just in the individual brain, but only part of it—will not help neuroscience in its quest for human consciousness.

ONE OF THE many occasions I corresponded with neuroscientists about their monist views on consciousness, an American professor (who shall remain anonymous) very kindly took the time to read my criticism and defended his position as follows:

"First, the theory is about the most basic form of consciousness and not about 'self' consciousness. As Seth discusses in his great book (Being You), those are different things.

Second, the theory is not about language-based thinking. Thinking in conscious creatures such as chimpanzees and very small children is not language based. As well, adult thinking is often not language-based (the Sapir-Whorf hypothesis was falsified long ago, at least its strong form was).

Third, humans with bilateral lesions of the hippocampus are perfectly conscious. Recall the famous case of H.M. They just can't have as conscious contents some episodic memories.

All the things you mention are interesting. The approach looks at simple cases because we do not have an understanding of how the brain creates these states, even the most basic forms of these states. When trying to figure out how cars work, it is better to study a Volkswagen beetle than, say, a Tesla."

Out of respect, and since the professor is a very busy man, I left it at that. The truth is—however—that, even though our mind, in its complexity, has managed to integrate two distinct layers, and operates seamlessly, travelling backwards and forwards between sentience and cognition, there is a clear separation between the two layers. As I said above, our biological brain is evolutionary, our human mind is meta-evolutionary. Any attempt at a gradual study from an amoeba to a human being will reach a very obvious ceiling. If the subject of our study is human consciousness, any evolutionary approach is clearly misguided.

The mainstream Neo-Darwinian scientific world attempts an explanation of consciousness following a gradual accumulation of random mutations through natural selection. That is wrong on two accounts:

1) Evolution can also occur through alternative mechanisms like spontaneous mutations, i.e., genetic insertions, or symbiogenesis;

2) Consciousness in humans has a cultural component that cannot be explained through physical evolution.

1) Gives an idea of how misguided the scientific world is; to appreciate this we only need to look at somebody like Richard Dawkins (with his emphasis on genes), and followers of the Modern Synthesis of Darwinian evolution. They believe that evolution is only due to the gradual accumulation of genetic variations, and that the creation of new species and taxonomic groups is the same, but due to extrapolations of events within populations on a larger scale.

Dawkins, supposedly a scientific leader, in his books *'The Selfish Gene', 'The Extended Phenotype'* and *'The Blind Watchmaker'* blatantly ignores the possibility of any type of evolution other than through random mutations.

Why can't biological evolution alone explain human consciousness? Are there any other indications as to why not?

Well, we all know that mutations mean change. Because of their randomness not all mutations are beneficial to an individual or to a species. Very often the result of a mutation is exactly the opposite. In those cases, individuals who have mutated die. What helps beneficial mutations is natural selection. Natural selection produces mutant individuals that are better suited to survive in a given environment. This is probably Darwin's crowning discovery. Random mutation needs to be combined with natural selection for the species to thrive.

Humans are the result of countless biological mutations that helped them survive up to a certain point. That point was the onset of humanity, the beginning of language and culture. Then, the mutations almost stopped being individual and became largely cultural. Certain cultures progressed further through beneficial mutations. Groups change and improve. These culturally-induced biological mutations should be easy to prove, I imagine. To learn to read, our brains need to adapt. After generations of reading, there must be areas that grow in individual brains. With cell phones, tablets and laptops, children are finding it increasingly difficult to read cursive writing. Are there any biological studies that can establish where in the brain that happens?

Biological mutations—which continue to occur to our species—are not helping individuals to survive any more. Human institutions provide against the survival of the biologically fittest. Human culture is the opposite of the "law of the jungle".

Human individuals belong in groups, and those groups provide support to individuals. There are rules that each culture imposes because groups without rules perish. The result is that rational individuals communicate within their own culture, and physical fitness, although desirable, is not necessary for the individual to survive and thrive in society. Intelligence is important, together with other qualities.

The emergence of human consciousness does not fit easily within Darwinian evolution. Wallace knew it. Darwin suspected it. There is no physical explanation. The only explanation is collective (cultural and dualistic). We cannot ignore the metaphysical nature of human consciousness.

THE MAIN REASON for the scientific bias towards physicalism and reductivism is that biology is a physical science and, as such, it is unable to find an explanation for collective consciousness, but there are other reasons that contribute to the rejection of dualism.

As we have seen, neuroscience has currently adopted a monistic approach. That has a logic to it. While Descartes, a philosopher, believed that human beings have a dual nature—what in those days was considered a body and a 'spirit'—, neuroscience has devoted itself to studying only the biological side of that nature and denies the existence of a 'spirit', or a 'soul'. Well, let's make it quite clear: nowadays few scientists believe that what drives human beings is a soul, but our mind (*psyche*), our consciousness. It obviously exists and it is intangible; but it does not emerge from neurones.

The undeniable news is this: humans are not born with our human capacities completely developed. We are born only equipped to be fully human. But at the time we can neither speak nor communicate. We do not understand our parents, our siblings, our relatives. They love us and care for us. And we can do nothing but allow them to love us and care for us. With speech and communication comes cognition; all of them are skills that humans need in order to live in society, and we are social beings. We need to live in society and we need society to live. A tiger lives on its own (yes, of course, it fights, hunts, and mates). We cannot live alone. We possess a highly adaptive capacity to learn from our experiences, but this capacity would not have survived prehistory without the cumulative generation-to-generation transmission of knowledge. Among humans, information is transmitted from one generation to the other, but also in space, contemporarily.

RIGHT NOW, there are many theories of consciousness but there is no universally agreed definition of consciousness. The trend is definitely monist, physicalist, and reductionist.

THE ASSOCIATION FOR THE SCIENTIFIC STUDY OF CONSCIOUSNESS held its 26th meeting at New York University from 22-25 June 2023. More than 20 theories of consciousness were discussed. There was no final winner, but two of them, Global Neuronal Workspace Theory (GNWT), presented by Stanislas Dehane, and Integrated Information Theory (IIT), presented by Sergio Tononi, were considered the two theories closest to the mark.

A significant highlight of the conference was a wager between Cristof Koch, a neuroscientist, and David Chalmers, a philosopher. Chalmers had challenged Koch to provide an explanation for the inner sensation of consciousness by examining the processes within the brain. Many philosophers remained sceptical of the idea that a purely physicalist approach could offer a satisfactory solution. The bet had a duration of twenty-five years, and with no discovery of neural correlates of consciousness during that time, Chalmers was declared the winner. The bet was renewed for another twenty-five years. By then maybe neither Chalmers nor Koch will be alive, but whoever the scientist is who is in charge of the wager, he or she will have to pay another box of port. There are no neural correlates where consciousness is generated.

The event showcased various perspectives on consciousness, with some theories emphasising the role of cognition, such as GNWT, and others taking a "back-of-the-brain" approach that did not account for cognition. The difference is that, if only sentience is considered—if what is being discussed is not "human consciousness"—then we are comparing apples and oranges. Depending on how you define consciousness, it is possible to say that many animal species are sentient, but only humans are conscious. More on that later.

Currently, there are several definitions of consciousness. Also, the perspective of the Cartesian dualistic framework (that recognises human nature as a combined physical and non-physical system) is now considered taboo.

Even though mind and matter are clearly two distinct ontological entities, only monist notions are currently acceptable. The majority of neuroscientists only study the brain as a physical, individual, entity. This paradigm appears to purposely ignore the existence of culture as a valid component of consciousness. Sadly, it is a solipsistic view of the world, where only the material, individual, aspects of the brain are considered 'mind'.

Some philosophers of science—like Alva Noë—however, reject monism and describe the *"human being as a kind of bio-cultural phenomenon"*. For what is worth, I totally agree.

Another central problem is that of feeling, or "qualia" if you like: the subjective phenomena that constitute biological experience. That is, how, as individuals, we feel pain, enjoy music, or go through other sensations. The question posed by Chalmers that originated the whole thing, the "hard problem of consciousness", relates to how you can explain qualia in terms of neural correlates in the brain.

The history of the whole happening began in 1990, when Crick —a Nobel Prize winner—and Koch, one of his students, published a paper, *"Towards a Neurobiological Theory of Consciousness"*, based on the notion that consciousness could be studied from a solely physicalist perspective within the brain. Ever since, the main goal of neuroscience has been to discover the neural correlates of consciousness, or the circuitry where consciousness resides. Some advances have been made in neuroscience and so far, the wager—a box of port—has been paid. Close, but no cigar.

From a lay person's perspective, the need for a clear and agreed-upon definition of consciousness is evident. To study human consciousness effectively, one must delineate its boundaries, which should encompass cognitive, cultural and linguistic aspects in order to fully include the 'human' aspect of that consciousness. A purely reductionist approach falls short of explaining the intricate exchanges and transformations of information that occur during cultural communication.

From an etymological point of view, objecting to the use of the term "consciousness" to mean "sentience" is easy. The term has Latin origins: "Scientia" means "knowledge" and "cum" means "with". "Conscientia", thus, implies "shared knowledge". The knowledge we share, socially or culturally.

Scientific reductionism attempts to circumscribe consciousness within a locus or loci in the individual brain. That is impossible: the term implies something intangible that allows communication with other human individuals. Using 'consciousness' interchangeably with 'sentience' needlessly isolates the individual and adds complexity to the discussion.

At this stage, in the words of a neuroscientist, neuroscience prefers to study *"a VW beetle"* in order to eventually be able to explain a *"Tesla"*. The analogy is not an apt one. In reality, neuroscientists would like to explain a rocket as those used by NASA, or Starlink by studying a Ford Model T. The evolution of the rocket involves the quantum leap taken by the Wright brothers: the invention of the airplane.

A REDUCTIONIST, Dr Kevin Morris, from Tulane University, author of *"Physicalism deconstructed"*, holds the view that the brain *is* consciousness. Apparently, there would be nothing over and above the physical world. That view is very difficult to understand for anyone who holds the commonsense view that ideas, and knowledge (and communication to a certain extent) are non-physical entities.

At an interview, however, Morris stated he believes in a more expansive concept of physicalism (?) than other philosophers: *"Physics, austere mathematical physics, describes things relationally, in terms of dispositions, in terms of how well one thing interacts with another, and you might think—for reasons that have nothing to do with consciousness— that there must be more to the world than just relations or dispositions."* That view appears to contradict the notion that the brain *is* consciousness. Things do not get much clearer from physicalists, other than anything metaphysical does not exist.

TO ADD TO THE CONFUSION, some neuroscientists are even negationists. Michael Egnor, a Professor of Neurosurgery and Pediatrics, for instance, believes that we should stop trying to understand consciousness because it is beyond what is humanly possible. He says so very clearly in a recent article:

"[Wittgenstein says] that there are things of such fundamental importance to us that efforts to speak of them inherently mislead

us. I believe that consciousness is such a thing, and that is why it cannot be defined. We cannot say clearly what we mean by consciousness, not because we haven't gotten the philosophy right or because we need more neuroscience experiments, but because conscious experience is too close to us, too fundamental to us to put in words. I think that Koch and, hopefully, other neuroscientists and philosophers are coming to understand this.

We cannot define consciousness or explain it by the methods of logic or neuroscience. Consciousness is that by which we perceive and understand, not that which can be understood."

From my perspective, things will become clearer: at a certain point, one of our hominid ancestors was intelligent enough to understand a message, a request—or probably an order—from another primate. He or she was able to connect a sound, emitted by the latter, with a meaning. He or she understood. That was all; that was the beginning of cognition within human consciousness. It was the beginning of humanity, which was cultural. And linguistic. The Adam and Eve phenomenon. The repetition of that act, probably through many generations, created neurones that ended up being gyri in the cerebral cortex.

It was a giant leap. After that moment, humans became the only animals with a complex, recursive, language that ended up involving present, but also past and future: i.e., possibility, long-term memory, and imagination, which other animals appear to lack or have in smaller degrees. Information became more complicated. Knowledge grew in sophistication. At the same time, the cerebral cortex expanded exponentially. Human brains needed larger craniums to contain them and women gave birth with pain. But above all, humans were able to co-operate to a much larger extent and excelled in functioning as social beings in large numbers.

Through religion, which is nothing but an earlier quest for knowledge—an old form of science—the ancient Hebrews explained how before language there was no humanity. I hinted above (as a non-religious person), that the myth of Adam and Eve in the Garden of Eden poetically addresses issues like communication, guilt, punishment, work and, basically, the beginning of consciousness, self-awareness and identity. Two hominins, one male and one female, become human. That was the myth. We are now aware that, from that moment on—not as God warns in the myth, but in reality—human lives would change dramatically. We learned to live in societies increasingly larger. And we developed something that was non-physical: a collective consciousness. A corpus of information shared by the collective.

What does not appear to be understood is that language and culture are a "black swan" phenomenon; the fact that an ape understood another ape was probably not that strange or not that far beyond normal biological evolution: other animals communicate. But if we deem that act as the onset of language, culture and society, it was definitely meta-evolutionary.

With the advent of humanity and cognition came other phenomena that were definitely not somatic, such as memory, imagination, communication, co-operation, creativity, and adventurousness, i.e., collective human consciousness. The latter is a non-physical entity that lives within a culture and interacts with the individual. A culture is more than the sum of the individuals that are part of that culture, it involves a history and a possible future, it is dynamic: it is synergetic.

Alva Noë (The Entanglement) succinctly describes what the science of consciousness needs to do now:

"Modern biology achieved its full explanatory power, its ability to account for life, its variety, and origins, thanks to the Original Synthesis, that is, the integration of Darwinian evolution with Mendelian genetics, but also with the new molecular biology that came of age in the mid-twentieth century. But if we are to explain the <u>human mind</u> (my underlining), it is now believed by many, we need a New Synthesis, that is, we need to join biology, so understood, to the theory of cultural evolution". Yes, totally agreed.

I repeat, human consciousness appears to have two layers; they are integrated, but they are definitely discrete, as they originate in different phenomena: one of them is evolutionary, and the other one, meta-evolutionary.

Yuval Noah Harari's account of scientific progress in the field of consciousness is devastatingly clear:

"To be frank, science knows surprisingly little about mind and consciousness. Current orthodoxy holds that consciousness is created by electrochemical reactions in the brain and that mental experiences fulfil some essential data-processing function.

However, nobody has any idea how a congeries of biochemical reactions and electrical currents in the brain creates the subjective experience of pain, anger or love. But as of 2016, we have no such explanation, and we had better be clear about that."

Eight years have passed and neuroscientists continue on the same path.

The title of Harari's second book (*Homo Deus*) suggests we are deities.

In essence, our species, *H. sapiens*, could be aptly named *Homo Creator*, as we have collectively forged an intangible entity—collective human consciousness—within our cultures and societies, which unites us as witnesses of the universe. Our religious ancestors believed in the Holy Spirit. Today, we have manifested a universal hologram of it in the digital realm.

A PHILOSOPHY OF CONSCIOUSNESS

"Finally, conscious experience, the method's source and touchstone, drops out of sight completely, hidden in the Blind Spot (the amnesia of experience). The result is a picture of reality drawn in physicalist, reductionist, and objectivist terms, from which consciousness has been excluded by construction."

- Frank, Gleiser & Thompson

The current scientific situation concerning the study of consciousness involves an obvious problem: as Western philosophy is totally individual-centred, based on solipsism, the proposed solutions to the problem of consciousness have always studied the individual. Human consciousness involves much more than the individual. Objective reality allows us to study objects outside of ourselves and excludes the individual from reality.

Frank et al, when dealing with what they call *'the explanatory gap'* discuss the point of separation from nature:

"The problem goes back to the rise of modern science in the seventeenth century, particularly to the bifurcation of nature, the division of nature into external, physical reality, conceived as mathematizable structure and dynamics, and subjective appearances, conceived as phenomenal qualities lodged inside the mind".

Not that it makes a great difference to the reality of the problem, but I believe it goes way back to the ancient Greeks, and to Saul of Tarsus' (aka St Paul) interpretation of Plato's immortal individual soul, which resulted in Christianity and the exaltation of the individual human being (who had an individual soul) above the rest of reality. The nature of the problem, then, is one that appears at the beginning of Western culture. Objective reality is born with the West.

IN *PHAEDRUS*, Socrates appears as introducing the allegory of the soul as a charioteer. The chariot has two horses: one of them is rational and the other one is passionate. The dual description of our mind could not have been more precise. Our consciousness is dual. We have what we now know as cognition and sentience. In his description, however, Socrates only concentrated on the individual. In this context, we need to remember that he also had clear ideas about the existence of a collective consciousness.

But even Descartes, who came up with the idea that mind and brain are not identical, based his theories solely on the individual. He maintained that the mind is indivisible and therefore, immaterial, whilst the body is exactly the opposite, divisible and material. This is the principle of substance dualism.

As emphasised above, the West, the ancient Greek philosophers, especially Plato (and Christianity, later) determined that we have a soul (*psyche*) and that that individual human soul, was immortal.

The way the argument started was with Plato's teacher, Socrates: he claimed that things that are perceptible—things that exist in space—are perishable; things that are intelligible, like ideas—which do not exist in space—are imperishable because they do not have parts and their substance does not decay. *Psyche* can understand those concepts, therefore, it is more akin to them, it is immortal like them.

They said: the body dies; the soul, which is the exact opposite of the body, does not die. In the case of Christian thought, the individual soul goes to Heaven (or maybe not). Well, we now think of *psyche* not as our ineffable soul, but as our mind, our consciousness (also ineffable?), and that is what we are trying to understand.

Descartes based his substance dualism on the fact that mind and body were completely different: their properties were quite distinct. They both existed but the nature of their existence was metaphysically dual and opposite. Descartes knew that his mind could think. His body could not think. Both, his body and his mind existed but they were separate entities, they were made of separate substances.

So far, within the study of consciousness, Western thought does not appear to contemplate the notion of a collective mind. Of course, people acknowledge the existence of the collective; the notions they have are clear, but science still believes the approaches to study consciousness should all be physical, begin with the individual, and be based on the brain.

AT THE BEGINNING of the book, I mention that in all probability we are the sole witnesses of the universe. That is what we do as human beings, but especially as Western individuals; that is what our consciousness means: a subjective witnessing of the reality that surrounds us.

The lone jaguar, the predator, sees the prey, but he does not witness, he acts; the starling flies in a murmuration with a thousand others and maybe enjoys dawn, or sunset, but he does not ponder that dawn or that sunset. From their non-subjective perspective, they are part of the scene. They cannot extricate themselves from reality.

Similarly, when a buddhist monk reaches *satori (*pure sentience), he empties his mind from cognition, he *becomes* reality in a human but unexplainable way; he witnesses holistically what is happening because he is part of that dynamics; like the jaguar and the starling, he becomes the activity. He is the garden and the flower and the arrow. He empirically understands reality, he 'lives' it. The Eastern experience, less solipsistic than ours, reaches its goal but it is an indescribable goal, unacceptable to the Western mind. It is something that the monk cannot explain. He can only demonstrate.

DAVID CHALMERS—THE dualist philosopher who won the bet against neuroscientist Cristof Koch last year—divides the situation into an easy problem and a hard problem. Cognitive sciences can solve the easy problem: for instance, how the mind can integrate information and control behaviour. What neuroscience cannot find are the neural correlates, or the exact place in the brain where the integration of that information is accompanied by experience, where consciousness emerges from neurones.

Chalmers questions how is it possible for any cognitive function to include experience. How do we know that an experience "feels" like that experience?, he asks. So far, as determined by the result of the wager at last year's meeting of the Association for the Scientific Study of Consciousness, there is no answer to that question: it remains a hard problem to solve.

Well, do we know what it would be like to experience the world without cognition, the way our hominin ancestors did? We can only speculate. Cognition and sentience are essential modules of the current conscious structure of our species. They are discrete but integrated, they cannot be peeled off. We have to understand how the system operates before we can begin to reverse-engineer it, before we can analyse the more basic components.

The limit of Western thought in terms of consciousness lies, then, where objective reality ends. We find that human consciousness can determine what kind of thing something is, but it cannot know what kind of thing *itself* is. The moment consciousness turns inwards, that knowledge becomes subjective, that reality is subjective.

ONE MAY or may not be interested in the views of Karl Popper in terms of philosophy of science; one may or may not totally agree with him. But his ideas on falsifiability and his taxonomical analysis of reality should be an important point of reference for the scientific world.

Popper's *"World 1"* includes all physical bodies plus radiation and gravity. Those are clear objects that science can study. He circumscribes his *"World 2"* to anything that occurs within the individual human mind (i.e., sentience and cognition); that is what is subjective; Popper's *"World 3"* include the objective elements of human culture, these are the sciences, lyrics, art, language, myths, stories, etc., that is, all the intangible, objective world we share as members of the human community. Human consciousness, I would say, mainly resides in *"Worlds 2 and 3"*, with a small physical component in *"World 1"*, that which is processed within the brain—the tiny electrical vibrations that occur when synapses transmit experience or thought from one neurone to another. As I state before, maybe we do not agree with the way Popper divides reality, but he was an expert, a philosopher of science, and one thing is clear from his perspective: science is not totally qualified to study human consciousness.

WE KNOW there are centres in the brain where some cognitive functions occur. We know the organ has centres for sentience, but explaining where the correlates are, where consciousness resides, appears to be impossible. Consciousness is basically something distinct from the body or any of its organs. The way it communicates can be perceived as something separate from the body.

Neuroscience can explain where some cognitive functions occur, it can explain where some sensations occur, but it cannot explain consciousness as a physical entity because it not a physical entity. It is not generated by neurones either. It is exactly the other way around: neurones are generated by it. Science cannot find an answer to the existence of consciousness because the answer lies outside its bounds.

The problem has a solution, though. What does this book propose? The solution is not only dual, but also hybrid. It should be approached from a different angle:

> **1. For starters, a clear definition of human consciousness is urgently required: should we study human consciousness, or just sentience, excluding cognition?—which is what neuroscience appears to be doing;**
>
> **2. A multidisciplinary approach is essential, but perhaps any consciousness research should be centred in the humanities;**
>
> **3. Human consciousness should be studied as a partially cultural, collective, social phenomenon that operates and is transmitted through biological means by the individual.**

Humans, as defined by Aristotle, are social animals. We function like birds or fish function in flocks or schools. We are not meant to operate in isolation, like certain other animals. We *need* the collective. That is how our species has developed post-biological-evolution thus far (and this doesn't meant that biological evolution has ever stopped). Any attempt to describe a human being will find itself confronted with society as well. When we talk about human circumstances, we talk about a certain period and a certain culture. Th study of human beings and their behaviour cannot be based on the individual.

The essence of human consciousness—I believe—as distinct from that of other animals, is easy to discern: the functions of the brain we are born with (all of them have to do with sentience); and the ones we acquire—high human consciousness, or *psyche*—, that develop with the intervention of the collective. We are born sentient. High consciousness is only acquired through parental and collective upbringing. It is culturally and individually transmitted. It is cultural by nature.

In order to understand how human consciousness operates, we have to imagine it within a certain society: clear evidence of it is that most psychological problems involve an individual that is dysfunctional, i.e., who cannot function, or who does not function properly, within his or her social environment.

Evolutionist studies of consciousness will always find themselves at a point where they cannot continue. The ceiling is located where biology ends and culture begins. The other problem is that both layers are integrated so, rather than discussing the dual nature of our consciousness, perhaps we should be discussing the *hybrid* nature of it. Because it is biological and individual, and at the same time it is immaterial and cultural. The immaterial, evanescent, component of our consciousness can only be studied philosophically, socially, culturally, metaphysically, i.e., only the humanities can address some of its elements. For instance, there is no way for the biological sciences to provide detailed explanations of semantics, creativity, adventurousness, or imagination. These are all hybrid phenomena that involve a great deal of cultural input.

Linguistics is a discipline to be found at the intersection of science and the humanities. We should take into account that language is biologically articulated and perceived, but it is still something produced by human consciousness at a cultural level. Language is not only a phonetic phenomenon, it is also a phonological one: i.e., we speak we articulate sounds, but at the same time we are communicating ideas.

Hominins only began communicating with each other in a more sophisticated manner, compared to other animals, once they grasped the meaning behind the sounds they were making. This moment, marked by the first phoneme, signifies the onset of humanity. Our evolution into humans didn't occur until one individual comprehended the message being conveyed by another individual.

ATTEMPTING to find consciousness in large language models (LLMs) is going to prove impossible for several reasons, one of them is that the LLM has no sentience, its mind is not hybrid like ours. It has all the appearance of a human mind, but it cannot feel because it was not born an animal, and there is no way you can explain, or transmit the concept of sensation or feeling to a non-feeling entity. The LLM is not alive.

What is a Large Language Model? They are the most complex, advanced type of language models. They are based on huge datasets, which appear to 'understand' and generate language (speech or text). Their input comes from transformers (a deep-learning model), internet, or artificial neural networks (models that use human and animal brain principles and structures). They have masses of information and can communicate via language. They are, however, neither human nor alive. Logically —unless science proves that things can be conscious—they cannot be conscious.

Another very important consideration as to why artificial intelligence (AI) will never become conscious is that, even though its algorithm appears to understand meaning, thus far it only skims through the surface of the semantic structure of language. There is no depth to its appearance of intelligence. It is indeed artificial, but I have serious doubt as the existence of any 'intelligence' in it.

Any AI system can solve problems that involve logic. Its algorithm is good at logic from a mathematical perspective. Of course, they are much faster at chess or *go* than humans. They can beat any human being at any given game they are playing, but I recently read something that is quite true: if there is a fire in the building, the human player will leave, whereas the AI will continue playing and burn. Also, AI systems cannot be compared to human intelligence; they are fundamentally different in that they deal with particular problems. ChatGPT and other LLMs are domain-specific. They are geared to solve specific problems, problems within one area. What is fantastic about them is that they use our language, which is such a large system (it can describe almost anything). The fact remains that AI systems require incredible amounts of human-generated information (which could be linguistic or visual) but remain constrained to one specific use.

There are now many theories called hallucination-confabulation theories to explain some inaccurate outputs that LLMs may generate. "Hallucination-confabulation" is, again, a misnomer. The terms would imply real intelligence when in reality there is none. LLMs do not hallucinate because they cannot imagine. They are only a digital expression of something based on a piece of text or source information in digital form.

A text includes ideas, feelings and sensations, but it's only the linguistic expression of those ideas, etc., thus, the LLM is a secondary iteration of the original idea, sensation or feeling, and a shallow one at that. The LLM is a human creation, a language model implanted on a machine that cannot understand how the collective (that created the language) operates. Thus far, the

LLM adopts the culture of the language that is input into it without anything close to comprehension.

New ambitious projects involve more transitioning from text-centred LLMs to models with multimodal "integration". The concept sounds good, but the fact remains that AI operates within a digital environment, which does not even come close to biological (electrochemical) functions. There will never be feeling where there is no life.

What we will need to accept about LLMs is that they are not going to remain just "stochastic parrots" forever. Even if they never develop feelings, they will eventually 'understand' the same way humans understand certain concepts. 'Understanding' means getting to know or represent in their 'mental' systems the key properties of a concept by capturing the relationship of those properties and the language that describes them. Once you develop a powerful representation of the concept, you have understood it. What the computer will never have is 'mental' agency. It will never think by itself, as it has no self.

Discussing, in a recent article, the current functionalist philosophy of the mind and the lack of mental agency in AI, philosopher and theologian Daniel Bentley Hart explains differences between human mind and algorithm:

"Neither computers nor brains are either syntactic or semantic engines; there are no such things. Syntax and semantics exist only as intentional structures, inalienably, in a hermeneutical rather than physical space, and then only as inseparable aspects of an already existing semiotic system. The functionalist notion that thought arises from semantics, and semantics from syntax, and syntax from purely physiological system of stimulus and response, is absolutely backwards. When one decomposes intentionality

and consciousness into their supposed semiotic constituents, and signs into their syntax, and syntax into physical functions, one is not reducing the phenomena of mind to their causal basis; rather, one is dissipating those phenomena into their ever more diffuse effects. Meaning is a top-down hierarchy of dependent relations, unified at its apex by intentional mind. This is the sole ontological ground of all those mental operations that a computer's functions can reflect, but never produce. Mind cannot arise from its own contingent consequences.".

Hart may come from a different angle, but what he says makes linguistic and philosophic sense. It is possible that, eventually, AI will 'understand' some meaning, but intelligence cannot be equated with mind or high consciousness.

NEUROSCIENCE DICTATES that human consciousness is located in the mind. The mind, according to it, lives within the nervous system—especially the brain—and determines how we think and act: i.e., what we do. When neuroscientists declare that the human mind is a collection of electrochemical impulses among neurons; they assume that human consciousness emerges from those neurones. Neurones have thousands of synapses that connect neurones with other neurones. According to neuroscience, that synaptic net—then—is the mind.

The current physicalist approach forces neuroscience into isolating the mind within the individual. That is clearly not true. One of the components of the human mind—or consciousness—is cognition, which is cultural. I use language to think, but I also use it to communicate.

For instance, I may be self-aware when I dream, but I have no identity. Neither do I have reasoning. Normally, dreams are irrational—they make no sense. They are a product of the individual mind, but they do not include any really cognitive function. They only reproduce experience devoid of reason. We ignore their ultimate function. We are rational only when we are conscious. We use cognition to communicate with other human beings. That is when we need to make sense. Mentally ill people are individuals who make no sense. They cannot function in society because they cannot communicate rationally.

I have not invented any of the words I'm using to write this, and neither has the reader. We are both speakers of a language and members a culture (in my case, an adopted one). Neurocientists, of course, accept interaction among humans as a fact. The difference is that they assume consciousness is generated within the individual. Can that be a correct assumption? What is generated within the individual is sentience, which includes experience, it includes self-awareness, but does not include identity, for instance, which is a cognitive-generated notion. Mothers—and fathers to some extent—are the ones that generate language and cognition. They only teach their babies and toddlers what they learnt from their parents.

ACCORDING TO JEWISH TRADITION, in the 16th Century, Judah Loew, the Rabbi of Prague, used mud from the river and combined different vowels and consonants until he could pronounce the only, the true, name of God. In doing so, he achieved the cabbalistic creation of a sort of living being, the Golem. The creature was meant to be a super human who would save the Jews. It did not happen. The Golem could barely sweep the floor of the synagogue. His brutish—human-created—soul barely allowed him to function. He could not talk.

Three centuries later, in 1816, Mary Shelley took part in an agreement, a competition, with Percy Shelley and Lord Byron to write the ultimate horror story. Mary ended up writing a famous novel about another monster, this time, one created by a gentleman, Dr Victor Frankenstein, who is a mad scientist. Dr Frankenstein is bent on creating an artificial man. The gentleman assembles the monster using body parts. And this creature is another failure.

We have discussed LLMs and other types of Artificial Intelligence. And, the way things currently are, it will be very difficult for them to acquire something similar to human consciousness as they operate from a very shallow basis (as we said, they are a secondary iteration of human discourse) and are not sentient.

The three examples above have similarities. The most important one is that the creations lack the equivalent of human high consciousness or, in the case of AI, any consciousness at all. They may understand some language input, but they do not actually perceive, they do not feel, they do not think the way a human being perceives, feels, and thinks. AI systems do not even perceive or feel like a living being. In biblical terms, I would say they lack the real "breath of life", what the ancient Hebrews called *"nephesh"*.

Human consciousness has evolved strangely, maybe because every individual begins from scratch. Every child learns everything subjectively. Every child is a new hard disk drive. As Borges noticed in *The Witness,* experience is forever lost every time someone dies. One good thing is that humans have notation systems and graphemes of all sorts. Thankfully, some thoughts—and feelings, as we saw in the case of Bashō—survive the individual.

THIS BRINGS us to a situation on the opposite side to consciousness and AI, that is, consciousness and the newborn. AI is not a live being and will never become alive or conscious, whereas the newborn is alive but has not acquired consciousness as yet.

WHILE ANALYSING prenatal and postnatal consciousness, a new joint study from several Australian, European and American universities—published in the journal T*rends in Cognitive Sciences*—appears to have discovered that babies become conscious even before birth. The study, of course, is based on current trends concerning the definition of consciousness (or lack thereof).

The study confirms that human consciousness is not immediate, but that it commences and grows gradually, almost imperceptibly. The markers used in the study were: functional connectivity, frontal brain networks, multi-sensory integration and neural markers of perceptual consciousness. They are behavioural and sensory.

I imagine newborn babies have sensory experiences from birth and would venture that foetuses may even have some prenatal ones. It is possible that newbies integrate sensory and some basic cognitive experiences to respond to their new environment. Even though there are some cognitive functions that may not involve language, I would be very sceptical to the existence of any prenatal cognitive function. There seems to be cognition without language, but I would find very difficult to believe that any form of cognition could be acquired without language.

IN AN INTERVIEW CONDUCTED some years ago, Neil deGrasse Tyson asked Heather Berlin, a neuroscientist and cognitive psychologist:

•*NdGT - Heather, what is consciousness?*

• *HB - ... We can define consciousness very simply as a first-person subjective experience. So, you are only aware that you have it. I don't know what your consciousness is like. I only know what it is from internally. So, what is consciousness? First person subjective experience. How is it tied to the brain? We're still trying to figure that out. Now, that's different from self-awareness. So you can be conscious without being self-aware...*

•*NdGT - Like in a coma or sleep?*

•*HB - For example, babies; they can be conscious, meaning you can have raw sensations like seeing the colour red or feeling something soft, or smelling a rose, without being aware of oneself, or having sort of metacognition, like thoughts about other thoughts. So, I'm going to say, I'm the one having these thoughts.*

•*NdGT - So if we're going to say another animal is not conscious... is not self-aware but has consciousness...?*

•*HB - Exactly.*

•*NdGT - ... the bee finds the flower and nature goes on.*

•HB - There are syndromes also where we see that people have an experience of being conscious, of experiencing things in the environment ... Correct?... without a notion of concrete self-awareness. Yeah, there are certain dissociative disorders, where people lose their sense of self, but they're still conscious, so consciousness is very unique, you don't need to have necessarily memory for it, you don't even need language.".

What happens during this interview is quite typical of the way scientists try to deal with the problem. First, Berlin affirms without the slightest hesitation: *"We can define consciousness very simply as a first-person subjective experience."* That is, human consciousness is solely individual. She goes on: *"How is it tied to the brain? We're still trying to figure that out."*

Then she says that it is possible to be conscious without being self-aware. I am sure she is not referring to human consciousness. If you are conscious, you are self-aware. One assertion contradicts the other. If consciousness is first-person subjective experience, then you need to be aware of yourself when you are conscious. Maybe she is confusing self-awareness with identity. Identity is only acquired through contact with the collective.

There cannot be any real development into the study of human consciousness unless Western scientists realise that their current approach makes no sense. Before any further advances are made, maybe the possible existence of the meta-evolutionary (layered, dual, hybrid) nature of human consciousness should be considered.

I have suffered from transient amnesia and have first-hand experience of the fact that you can be conscious without knowing who you are; but you are self-aware, otherwise you would not be conscious.

A PHILOSOPHY OF CONSCIOUSNESS

What makes matters even more complicated—and I am not going to refer to free will, which physicalist scientists do not want to acknowledge—is that identity and self-awareness have very different natures. Self-awareness is biological and present, whereas identity is hybrid and time-related. You are self-aware the moment you are conscious, even when you are alone. Self-awareness occurs only in the present; you cannot be self-aware in the past or in the future. You can remember or imagine being self-aware, though.

Identity, on the other hand, is hybrid because you acquire your identity through the collective. A tiger cannot have an identity. Maybe elephants or whales have a semblance of identity because they live in groups. We don't know. Human beings have identities because identities are needed to function within their social groups. The time-related aspect of one's identity has to do with the continuity of that identity through the name society gives the individual. Identity is permanent. For the individual and for society. It may vary in the case of gender change, but that is only a socially accepted exception.

Here we cannot help but returning to Heraclitus, the river is the same and the individual is the same. How? To some extent — they are both different, pretty much like Theseus' ship: some elements may have gradually changed. The secret of the continuity is the day-by-day imperceptibility of that change.

Anil Seth (*Being you*) concurs with other neuroscientists; he begins one of the chapters of his book:

"It seems as though the self—your self—is the 'thing' that does the perceiving. But this is not how things are. The self is another perception, another controlled hallucination, though of a very special kind. From the sense of personal identity—like being a scientist, or a son—to experiences of having a body, and of simply

'being' a body, the many and varied elements of selfhood are Bayesian best guesses, designed by evolution to keep you alive."

What is noticeable here is that Seth does not appear to differentiate between 'identity' and 'self-awareness'. Maybe he tends to ignore that there is a difference because it conveniently supports his approach to consciousness: *"...being a scientist, or a son..."* involves much more than self-awareness. It involves—to a much greater extent—the perception of society rather than that of the individual. Evolution may have designed 'selfhood' to keep you alive, but it has not designed 'identity': that is a social creation in which the individual participates.

Seth gives a perfunctory nod to identity and its place in the social network:

"All these ways of being-a-self can be in place prior to any concept of personal identity—the identity that can be associated with a name, a history, and a future... for personal identity to exist, there has to be a personalised prior history, a thread of autobiographical memories, a remembered past and a projected future." And he goes on to say: *"The social self is all about how I perceive others perceiving me. It is the part of me that arises from my being embedded in a social network. The social self emerges gradually during childhood and continues to evolve throughout life..."* None of it explains how all of this can occur within a purely physicalist framework of human consciousness in the individual brain.

In any case, this brings us back to the main problem we discuss in this chapter. In their study of consciousness, the scientific community seem to disregard the existence of the collective. They affirm quite clearly that human consciousness does not extend to intersubjectivity. What they are saying does not make any sense.

THE EXPERIENCE of being oneself is also discussed by Frank et al *(The Blind Spot)*:

"The sense of self is not one thing. It includes many different elements, such as the bodily self, the mental self experienced in memory and anticipation, the self as a personal storyline (the narrative self), and the social self.".

Bodily self-awareness is something that only happens in the present and is part of sentience. The mental self as experienced in memory and anticipation (past and future) is obviously part of high consciousness, as time is only a human construct. Animals have neither long-term memory nor anticipation (at least they do not appear to imagine long-term into the future). There are many cases of dogs that keep on waiting for their humans to return because they cannot imagine the possibility of a permanent absence. Like the previous one, the self as a personal storyline, should actually be considered identity, as it also happens in time and within the collective, and so is the social self, only that the latter one is a spatial self.

ONE FINAL POINT in this chapter. Again, Frank et al, argue that consciousness has primacy over knowledge, i.e., that knowledge is included within consciousness.

"Claiming that consciousness can be reductively explained by its abstract structural residues in physics or cognitive neuroscience inverts the whole procedure of generating scientific knowledge, which starts from and is forever beholden to direct experience. The move is absurd in principle because it tries to replace the subjec-

tivity it excluded at the start with an abstraction cast in completely objective terms. It fails to recognize the ineliminable primacy of consciousness in knowledge.".

I agree with what they are saying in terms of objective reality. It is kind of clear, but how does it happen? Somehow,—it seems to me—the idea of explaining human consciousness from a chronological perspective appears to provide a clearer picture: humanity originated in what Frank et al call the life-world, that is, our first ancestors were very basic living creatures; then they became sentient mammals; then they became one of the primate species; then language appeared among those primates —which constituted a meta-evolutionary phenomenon and added the extra layer of cognition over sentience—; then human groups grew into cultures and they, in turn, grew into civilisations. In this explanation, knowledge is acquired and transmitted through culture.

Cognition, then, is subsumed within human consciousness, it becomes a human addition, a layer over our exclusively-animal-consciousness, but the analysis—I believe—is clearer by means of the chronological explanation. To borrow from (microcosmic) existentialist terminology, *existence* precedes *essence*. I would add, *we are* [biologically animal] before *we become* [culturally human]. What is interesting in all of this is that, chronologically, it applies to the individual as well as the collective. There is also a circularity to it: a human individual initiated culture; and culture keeps on creating human individuals.

LANGUAGE, CULTURE AND CONSCIOUSNESS

"A great stride in the development of the intellect will have followed, as soon as the half-art and half-instinct of language came into use; for the continued use of language will have reacted on the brain and produced an inherited effect; and this again will have reacted on the improvement of language. As Mr. Chauncey Wright has well remarked, the largeness of the brain in man relatively to his body, compared with the lower animals, may be attributed in chief part to the early use of some simple form of language,- that wonderful engine which affixes signs to all sorts of objects and qualities, and excites trains of thought which would never arise from the mere impression of the senses, or if they did arise could not be followed out. The higher intellectual powers of man, such as those of ratiocination, abstraction, self-consciousness, &c., probably follow from the continued improvement and exercise of the other mental faculties.."

- Charles Darwin, *Descent of Man*

. . .

*I*n *Descent of Man*, it is possible to appreciate that Darwin himself could see the effect language (and culture, of course) had had on the human brain. He could see the feedback-loop that had acted on the adaptive brain and produced the growth of the neocortex. Darwin guessed that—at a certain point in time—there had been a massive leap (meta-evolutionary—I would add) that placed *H. sapiens* well beyond sentience and any other species.

It is not that cognition is entirely dependent upon language, but it could be said that there is a pre-linguistic cognition and a linguistic one. Their natures are totally different: one appeared as a totally subjective phenomenon, whereas the other one emerged as an intersubjective explosion of knowledge that continues to grow to this day.

Our minds operate at two different levels to achieve two different ends.

One layer of the mind—sentience—is there to keep our individual bodies alive. Like all other mammalians, we need our senses to see, smell, hear, taste and touch. They allow us to move unimpeded in our surroundings, enjoy food or music; see, smell or hear predators, enemies, or sexual partners; and recognise familiar shapes and textures with our fingers or toes when we cannot see them, among other vital activities we require to survive as individuals. That layer is totally related to biology, totally physical. It is a result of evolution and all mammalians have it.

The biological part of our consciousness, which is also the 'emotional' one, concentrates mostly on survival, it regulates body temperature, breathing and heartbeat, for instance. That is the animal (as opposed to human) component. Whenever there is trauma, or danger, the mind initiates its response: fight or flight. All kinds of chemicals flood our body, from adrenaline to cortisol. Rational thought is absent. The subconscious is also at play, as what happens in dreams.

The main concern of this book is the second part—human consciousness—, which includes cognition and metacognition, and is mainly geared towards the gathering, processing, and communication of information. The gathering of information is accomplished either through hearing, vision or touch. The processing of information occurs within the individual and is—in the words of George Steiner: completely "*impalpable*". The third function, i.e., the communication of information is carried out either verbally or visually encoded (to be decoded by the recipient of that information, as in reading).

Those three functions, then, are hybrid, in that they require some physical input, processing and output. But there is a difference: the difference lies in that two of them—gathering (or receiving) and communicating information—involve external means, (whereas retrieving from memory and processing are solely individual). The first two functions are mainly linguistic / cultural.

The physicalist study of the human mind occupies itself with sentience and—to some extent—one third of human consciousness, i.e., information processing, which is carried out by the individual. It blatantly ignores, or attempts to ignore the other two functions of consciousness: gathering and communication of information.

At the beginning of this chapter, we saw that Charles Darwin had guessed that language greatly influenced cognition. He could surmise that dualism provided a good explanation for consciousness. Unfortunately, the nineteenth century was a period when the only option he had was materialism or religion. With no other rational possibility, he chose materialism. He often repeated that human beings were too proud to believe anything but Creation: *"Man in his arrogance thinks himself a great work, worthy the interposition of a deity, more humble & I believe true to consider him created from animals"*.

His biographers—Adrian Desmond and James Moore—quote him as he adopts an unbelievable explanation for consciousness, only understandable in a person of Darwin's intellect because of the times he lived in:

"[Professor John Elliotson's] stock provocation was that the brain exudes thought as the liver does bile. It was Darwin's <u>bon mot</u> exactly. 'Thought, however unintelligible it may be, seems as much function of organ, as bile of liver'. But Darwin's goading had a sting that even Elliotson's lacked. Everyone accepted that gravity was an intrinsic 'property of matter,' no one made a spiritual adjunct. So 'Why is thought' not seen as 'a secretion of [the] brain' in the same way? 'It is [because of] our arrogance, it is our admiration of ourselves'.".

What neuroscience proposes nowadays—not in the nineteenth century—is a very similar proposition: consciousness emerges from neurones.

~

IN HER BOOK *Kindred - Neanderthal love, death and art*, Rebecca Wragg-Sykes discusses the possible existence of Neanderthal cognition and communication abilities, whether Neanderthals could think and speak like us:

"Feeling wonder inside oneself is one thing. Being able to share an awe-inspiring or transcendent experience is much more powerful. For a meta-physical life to emerge, language is key because it allows emotions and meaning to become crystallised. Whether Neanderthals had any kind of language is, of course, one of the most enduring questions about them."

Wragg-Sykes then compares their brains, ears and throats with that of *H. sapiens*: *"Neanderthals flatter foreheads had less room for the frontal cortex area, intimately connected to complex thought processes like memory and language. ... today it appears that Neanderthal throats could make pretty much the same range of sounds as ours."*, and then decides that *"If this anatomy in humans is regarded as specialised for language, then Neanderthals could not have been so different."*

In the Introduction we saw that other species, like macaques, may have not only sentience, but a small degree of cognition as well. For the purposes of this chapter, the most important part of Wragg-Sykes' comparison is the impact of language in human consciousness in *H. Sapiens*, *"...for a meta-physical life to emerge, language is key"*. Totally. We have language.

LANGUAGE AND COGNITION appear to develop almost simultaneously. In fact, we don't know whether there is actual human cognition without any language. Indeed, it would be very difficult to imagine humanity without language.

Some written languages are so finely tuned that they allow thought be transmitted directly, although they usually have some phonetic component. Chinese and Japanese ideograms—*kanji* in Japanese, *hanzi* in Chinese—may convey ideas without the use of sound. The idea can be visually processed directly by the brain. For instance, 竹 means "bamboo" in both, Chinese and Japanese. In Japanese it's pronounced [ta:ke]; in Chinese it is [tzūdzə] with slight tone variations from Cantonese to Mandarin. The symbol is the same. These graphemes, these ideograms, are actually units of meaning. They don't represent individual sounds.

Of course, humans can think without producing any vocal sounds. That is called "inner speech". Some toddlers tend to vocalise their thoughts for a while, until they discover they don't need to do it. Most humans have that inner voice, which helps them articulate thought more clearly in their mind. The strength of that articulation is such, that often—especially while reading—the larynx produces minuscule reflex movements that accompany the inner voice.

If we think literacy commenced millennia ago, reading without making any sound is a fairly recent innovation. At a time when people would only read aloud, St Augustine marvelled at St Ambrose reading in silence:

"But when he read, his eyes glided through the pages and his heart searched for meaning, but his voice and tongue rested. Often, when we went to see him (because no one was forbidden to enter, nor did he want anyone to be announced to come to see him), we would see him like this, reading to himself...". Someone, in silence, acquires information from symbols. Quite remarkable. Augustine cannot stop wondering about his own consciousness.

"How come I have consciousness"—Augustine says—*"and animals don't?" "I turn to myself and ask myself, 'Who are you?' and 'One Man' will answer. And I find that in me there is a soul, and a body; one outside, and the other, inside me. ... Animals,*

large and small, can see the body, but they cannot be asked, because they have no use of reason besides their senses to judge what they see. Men can do it..." What's interesting is that Augustine sees that he is partly his body, and that his soul is "inside" that body. In other words, he places his self-awareness in an intermediate place between body and soul. But then Augustine lives in a period when psyche is not consciousness but a soul given to us by God.

IN THIS CHAPTER we are discussing language and culture. Apart from gesturing—which often implies language, or is included in language—what is involved in communication between human individuals is a mutually intelligible language. That means that, to communicate with a certain degree of sophistication, individuals need to belong to the same culture and use the same language.

In the Preface, I make reference to two previous books on the subject where I maintain, and I still believe, that human consciousness involves two layers and that psyche—the exclusively human component—is culturally transmitted. Saying that has consequences, implications and ramifications in all kinds of fields. *Inter alia*, it tends to support the Sapir-Whorf Hypothesis, or Linguistic Relativity.

Of course, many linguists do not agree with it. Recently, researchers have tested different versions of the hypothesis in the field of neuroscience behind language and human communication. They ended up identifying a 'universal language network' in the brain. They affirm there is a genetically predetermined structural neural network. Despite huge differences in their languages, subjects of experiments demonstrated that key properties in their brains' language network were consistent with each other. That would not really falsify linguistic relativity, I would say.

Anthropologists and linguists in the field are constantly finding cultural differences that affect the way speakers of certain languages perceive the world.

Remember the movie "Dances with wolves"? That was the name the Sioux had given the protagonist. That name would not have occurred to an English native speaker. It's a sentence. It sounds strange. English has a predominance of nouns (it is a more static language), whereas languages indigenous to the Americas, like Arapaho, place the emphasis on verbs (they are more dynamic than their European counterparts). The word for 'cement' in Arapaho is roughly 'it has hardened', and the term for 'chair' is 'the place where you sit'. In these two cases the language emphasises function rather than feature. It is something that happens instead of a static characteristic of the subject. These differences permeate languages and inform the way a speaker sees the world.

BUT LET'S revisit the Linguistic Relativity hypothesis. In the early twentieth century, Edward Sapir was a teacher of Anthropology and Linguistics at Yale University. He believed that *"Language is a purely human and non-instinctive method of communicating ideas, emotions and desires by means of a system of voluntarily produced symbols"*.

Benjamin Lee Whorf, one of Sapir's students at Yale, believed language and culture were directly involved in the evolution of thought processes; he also believed that Western science ignored the differences caused by the phenomenon and that—consequently—Western scientific assumptions were unnecessarily narrow because they relied solely on a Western logical system.

Sapir and Whorf worked together on the hypothesis that the grammatical and verbal structure of a person's mother tongue influences that person's perception of the world. The idea, originally known as Linguistic Relativity, was fiercely rejected by universalists. It still is.

Coincidentally, that notion tallies with one of Whorf's studies that was also rejected as not a true reflection of reality: Whorf stated that the Hopi, a native American nation, lacked the concept of time in their language.

If time is a human construct—as I, among others, claim—that developed within human culture, or a device that we use to explain our long-term memory, time exists only within human culture. Depending on their need, it is possible for some cultures not to have developed the concept of time altogether, or to have developed a partial, or different, concept of time from what we consider "normal" in the West.

We experience and interpret reality the way we do because we are predisposed by our language and by the way our culture perceives it.

Similarly, some Australian aboriginal languages include totally different ideas concerning pronouns, and times and modes of verbs.

The Pirahã, a fairly isolated Amazonian tribe, are probably a good example of linguistic relativity. The Pirahã language, like some Melanesian ones, lacks cardinal numbers after "one" and "two", even though the Pirahã tribe understand larger quantities; it has no colours except "light" and "dark" (it has other ways of explaining colours: "like-blood", for red); and it includes a system of pronouns—that can also become nouns—that is extremely difficult to understand by Westerners.

To give you an idea of the difficulty involved in understanding Pirahã grammar, their verbal system has a quantity of aspects: perfective (completed), imperfective (incomplete), telic (reaching a goal), atelic, repeated, and commencing; but they have very little transitivity. It appears fairly clear that the way the Pirahã language has developed is a reflection of the way they perceive the world. Languages and cultures develop according to the needs of a particular society in a given environment.

Another culture that disproves the complete universality of language are the Amondawa, also another fairly uncontacted Amazonian tribe. They do not have a word for "time". When asked to find an equivalent to the Portuguese word for time, "*tempo*", they came up with their word for "sun". They don't appear to conceive of time as a flowing, measurable, framework for change. These examples may eventually prove that the notion of time is shaped by culture. Coincidentally, speakers of Kuuk Thayorre, in Queensland, Australia, when faced with some temporal progressions (e.g., same man, different ages), tend to follow, in their thinking, the trajectory of the sun.

But let us go back to the use of language as the main means of communicating information. The current insistence of neuroscience on the assumption that human consciousness emerges from the neurones in our brains blatantly ignores two thirds of how our brain functions. We were saying before that human cognition includes three information functions. Two of them are communicative: input and output of information are generally carried out as part of linguistic and cultural information exchange with other human beings. Yes, we may discover facts by ourselves, but normally, we learn through input coming from other individuals.

Humans can imagine things that do not exist, and they can do it individually or collectively. Yuval Noah Harari adds that human language has an extraordinary quality that is lacking in any other form of animal communication: apart from imagining non-existent things individually, humans can share those ideas and express them collectively.

Humans can add an extra layer to reality: they can create a special social reality and invent the rules that apply to that collective. They are the constitutive rules that are understood and respected by all members of that society. From the time we are toddlers, humans are taught increasingly complex systems of rules that we will need to abide by in order to live within society. Those complex, intangible concepts are transmitted only because human language is capable of expressing and sharing them.

A REDUCTIONIST FRAMEWORK fails to account for the intricate transformation of information involved in the communication process. For example, when describing pain to a medical professional, one must perform several mental and physical steps:

a) Recollect previous instances of the same physical pain (in a non-physical manner), make comparisons, and consider its location. b) Think about the pain and formulate a description using words, typically in a language understood by the doctor. c) Articulate these words using various anatomical components such as the larynx, vocal cords, mouth (including teeth, tongue, palate, glottis, and lips), lungs, and nose.

Once these meaningful sounds, or phonemes, are uttered and reach the doctor's ear, the doctor must engage in a reception process that essentially reverses the emission of information. This process involves interpreting the sounds to decode the conveyed information. Sorry, describing such a cultural exchange of information is kind of complicated.

Within the process I have described above there are biological phenomena (the somatic segments of the linguistic process), non-biological linguistic phenomena, and purely cultural phenomena.

George Steiner elegantly explains the complex nature of the phenomenon:

"Language is assuredly material in that it requires the play of muscle and vocal cords; but it is also impalpable and, by virtue of inscription and remembrance, free of time, though moving in temporal flow".

So, the way things are, neuroscientists are looking for the correlates where sentience and cognition are generated by neurones. They want to find out where sentience and cognition become experience. How do we know that we feel something? How do we know that we know something? Their search leads nowhere because the question they formulate is the wrong one. Neurones are created by experience, not the other way around, and sentience is biological, whereas cognition is cultural, just as language is cultural.

Sentience and cognition have developed in such a way that they are distinct but integrated layers. The layers cannot be peeled away.

LANGUAGE, CULTURE AND CONSCIOUSNESS

An April 2024 article written by Coral del Val, *AI Finds Personality Shapes Genes*, in the eMagazine *Neuroscience*, reports on a study conducted by specialists in genetics, medicine, psychology and computer science from several Andalusian institutes of high learning, jointly with Washington University in St Louis, Baylor College of Medicine (Texas) and the Young Finns Study (Finland). The study was published in *Molecular Psychiatry*.

Using AI, the researchers analysed data from the Young Finns Study (analysing three levels of self-awareness in individuals) and their results indicated that there was *"a network of 4,000 genes that clustered into multiple modules that were expressed in specific regions of the brain. Some of these genes had already been linked in previous studies to the inheritance of human personality"*.

More importantly for the purposes of the hypothesis of this book, these genes were subdivided into two sub-networks:

"One network regulated emotional reactivity (anxiety, fear, etc.), while the other regulated what a person perceives as meaningful (e.g., production of concepts and language).".

The findings appear to support the ideas behind this book: 'two discrete but intertwined layers (*"networks of genes expressed in the brain"*) of human consciousness': a sentient, biological one (*"anxiety, fear, etc."*) and a cognitive, meta-evolutionary one (*"e.g., production of concepts and language"*).

RAFAEL PINTOS-LÓPEZ

LINGUISTS KNOW that only in rare instances there is real, direct, correspondence between two or more languages. Translators and interpreters, of course, are even more aware of that because they experience it in their day-to-day work. Those differences appear at all grammatical levels.

Of course, it is possible to look for a universal origin of language or state that we are born with a *"tabula rasa"* template in our brains that allows us to input and use any language. Nobody denies that we can learn any language. Of course, we can understand other cultures. The shared humanity of all races is a fact with which linguistic relativity is not concerned.

But also, what is impossible to deny is that culture and language are inextricably intertwined. The translation of a concept may involve a philosophical worldview that does not occur in the target language. In those cases, the expressions are really untranslatable. At least, when they are translated, they do not include every nuance that the interlocutor is meant to receive.

A good example would be どぞよろしく (dozou yoroshiku), in Japanese. In general, that means "pleased to meet you", so it can be translated with those exact words, but the actual meaning is "please take care of me" or "please, treat me favourably". A more formal way to say it in Japanese would be どぞよろしくおねがいします (dozou yoroshiku onegai shimasu), which can be translated as "how do you do?". But it actually means "I am so grateful for any support you may provide to me" or words to that effect. "Please" is included twice in the sentence, in different forms. English speakers find the concepts hard to understand. "How do you do?" sounds much better. That is how translation works, or appears to work. When that happens, the translator or interpreter knows that the semantic cultural transfer has not actually occurred, i.e., the actual meaning, the high degree of respect to the other person implied, has not come through (and cannot come through, or be understood, when the interlocutor is monolingual).

A Western mind finds that kind of respect almost impossible to understand. The other side of the coin is that an Asian mind finds the Western individual extremely solipsistic, to the point of being often culturally unacceptable. To the Asian ear, Westerners often sound rude. There are historical, philosophical and religious reasons for the difference, but that is another story.

The example I have just provided is an extreme one. There are countless examples like that one, though.

Differences between languages within the European language family can be important and may include instances that are impossible to translate correctly. For instance, Spanish divides space three ways (incidentally, like Japanese) : relative adjectives are "este" ("éste" for the pronoun), "ese" (ése), and "aquel" (aquél), and are marked by gender as well (feminine "esta", "esa" and "aquella"), whereas English has two divisions: "this" and "that". There is no adjective or pronoun for "that one over there", which is far from the speaker and the interlocutor as well; neither is there a gender marker.

But one example that gives a clear idea of worldview is when the speaker indicates that they have dropped something. In English it is "I dropped the cup". In Spanish it becomes "Se me cayó la taza" ("The cup dropped off me"), where responsibility for the cup being dropped rests solely on the cup.

Malagasy language and culture are, again, very difficult to explain to a Westerner. Malagasy, the official language of Madagascar, belongs to the Malayo-Polynesian family of languages.

One of the most striking features of Malagasy is that it does not include the idea of "left" or "right". It is what linguists call a 'spatially absolute language'. Western languages' speakers—born within egocentric cultures—find that difficult to comprehend, because whenever we talk about space, we do it by relating it to ourselves. Malagasy speakers do not think in terms of body-relative space, their absolute space is based on the cardinal points, and that type of orientation is only accomplished through a deep spatial sense which is, in turn, acquired from the culture. The other very interesting feature of spatially absolute languages is that their speakers seem to structure their memories and their general conception of time according to it.

If you are driving and, at an intersection, you ask one of the passengers whether to go left or right, they will not understand you. They will answer "yes" to avoid any possible confrontation, but will not give you the information you need. If you decide to turn left and that is the wrong direction, the interlocutor will tell you, "It would be better for you to go a bit further North", meaning that you should go in the other direction.

Giving directions like "left" or "right" would put the speaker at the centre of attention. That is not acceptable. Even within a dwelling, a Malagasy speaker would tell you that you can find the glass you need "to the South of your hand".

As strange as all of that sounds from our point of view, their language is not an exception. There are several languages and dialects that only use cardinal points to give directions.

The idiosyncrasies and complexities of Australian Aboriginal languages like Dyirbal, Kuuk Thaayorre, or Guugu-Yimidhirr, spoken in Northern Queensland, are almost incomprehensible without having lived in their communities. For instance, in both, Dyirbal and Guugu-Yimidhirr, there are special language registers—i.e., levels of respect or formality—even separate lexicons, to be used within different social contexts. No Guugu-Yimidhirr male speaker may address his mother-in-law or even look at her; and he has to use a different vocabulary to speak to his brother-in-law and keep a certain distance. He cannot look him in the eyes either. The culture divides society in *moieties* and the speaker of the language has to respect that division. Some of the relatives of a male speaker are *dhabul* (taboo). Addressing any of them has to be done in special ways. This is just a small sample of the many idiosyncrasies of Australian Aboriginal languages that indicate a clear difference in terms of the culture and the thought processes of their speakers. Communication between one of them and a speaker of a Western language is often difficult and fraught with misunderstanding.

STUDIES OF CULTURAL/LINGUISTIC conditioning that confirm Linguistic Relativity are beginning to be conducted more often, as researchers commence to accept the possibility of a non-universalist framework for the study of consciousness. Any monolingual person, on encountering a different culture —when they find themselves among people who speak a different language— often experience what is known as "culture shock". The characteristics of their language involve another perspective on the world. Humboldt used to say that different languages reflect different worldviews, that language forms thought. And, indeed, there is a loop: words shape thoughts and vice versa.

Recent research conducted at the University of Miami determined that the perception of pain in bilingual individuals may vary depending on the language they use at the time of expressing the pain or their inclination towards the Hispanic or Anglo cultures.

The study is described in a January 2024 article in *PsyPost*, *"Neuroscience study reveals how language affects pain processing among bilingual individuals"*, by Eric W Nolan. The author states:

"Interestingly, a participant's cultural orientation—whether they identified more with Hispanic or US-American culture—played a significant role in how language influenced their pain perception. The pain response was stronger in the language that matched the participant's stronger cultural orientation. This suggests that cultural identity can modulate the way language impacts our sensory experiences, including pain."

What is fascinating in this study is that participants felt that the pain was more intense when they spoke Spanish than when they spoke English. But the ratings were not just subjective ones. The language activity of the subjects was measured through MRI scans. These showed that the brain areas where pain is processed were really more active in the Hispanic context. Language and culture, then, appear to influence not only cognition, but sentience as well.

AND WHEN DID we acquire human culture? Again, let's see what Wragg-Sykes comparisons with our *Neanderthal* ancestors tell us about *H. sapiens* culture. She wonders: *"Could there really have been a 'light-bulb moment' when some novel genetic*

mutation or combination greatly increased H. sapiens tendencies towards more formalised artistic traditions, or flashy burials? Again, the reality is inconveniently uncertain."

Her assumption about *"some novel genetic mutation"* appears to be close to what neuroscientists believe. The reality may be uncertain, but common sense seems to dictate that *H. sapiens* were further advanced in the long process of the acquisition of language. When compared to *Neanderthals, H. sapiens*, who lived in larger groups, appears to have had a more developed (maybe recursive?) language that allowed for a more elaborate artistic expression and treatment of the dead.

A FEW MILLENNIA after Neanderthals had become extinct, during the Upper Palaeolithic, from about 34,000 to 24,000 years ago, the Gravettians, a hunter-gatherer pan-European culture were thought, for a long time, to be homogeneous throughout their territory. There was no Pan-European culture. Recent studies of their art and ornaments suggest that there were clear divisions: more than nine distinct groups are apparently discernible. What is interesting is that the analysis indicates that their ornaments communicated meaning. Affiliation to a group or social rank were evident from the beads these hunter-gatherers wore. But rather than geographic or ethnic differences, the message was cultural: *"I belong to such and such a group"*. The ethnic origin or place where the person lived were not as important as the group. The cultural message is still important today. It is possible that these groups fought with each other, but they also traded and interbred.

The Gravettians were faced with extremely inclement weather during the last Ice Age and were eventually replaced by other hunter-gatherer groups and other cultures. The prehistoric cultural and linguistic layers changed, moved and were replaced until they became recognisable civilisations like the Creto-Mycenaean one in what is now Greece, for instance.

∽

HUMBOLDT THOUGHT that literature led to the essence of a language. This, he believed, was the "spirit of the language". Linguistics, however, developed basically by studying language in terms of form rather than depth, rather than essence. In a way, Humboldt's ideas were more concerned with the way some aspects of a culture coincide with the development of its language and vice versa. Those fields are now more concentrated in fields like Ethnolinguistics and Semantics.

During the twentieth century, linguists generally believed that their discipline had to be considered as a serious (hard?) science, rather than something closer to the humanities. The way language may influence cognition and the way culture may influence language were taken as foggy areas that were outside their field. Chomsky and his view of linguistics and science were the ideal perspective for the times. It was the right moment.

The argument had commenced with Franz Boas, a nineteen-century anthropologist who was extremely interested in the links between consciousness, language and culture. Being interested in those links was considered pretty close to being a declared racist. We cannot forget that in those days Western Europe was still colonising Africa and natives were kept at the very bottom of the hierarchy. People mostly believed that their cognitive abilities were to blame for their situation. Towards the beginning of the twentieth century, the conflict was resolved by deciding that there were no racial differences in terms of consciousness. All human beings were the same. When Sapir and Whorf came up with their hypothesis, the whole thing was suspiciously close to ideas that had been proved wrong.

Later, Chomsky declared there were no structural differences between languages. If any dissimilarity existed, it was superficial. Humans were born with a linguistic template that was universal. Language and cognition were separate entities. There was no place in Linguistics for relativity.

The problem is that the study of language should be the main gate towards a deeper understanding of consciousness. Linguistics should not be considered solely a reductive, analytical discipline. It should be a hybrid discipline because it partakes of the richness and complexity of the humanities.

INTERSECTION WITH BIOLOGY

*I*n his book *Being You*, Anil Seth explains that one of the main objections he finds with substance dualism is that there is no explanation as to how the physical interacts with the non-physical.

"Sitting awkwardly in the middle [between physicalism and idealism] <u>dualists</u> like Descartes believe that consciousness (mind) and physical matter are separate substances or modes of existence, raising the tricky problem of how they ever interact. Nowadays few philosophers or scientists would explicitly sign up for this view. But for many people, at least in the West, dualism remains beguiling. The seductive intuition that conscious experiences seem non-physical encourages a 'naive dualism' where this 'seeming' drives beliefs about how things actually are. As we'll see throughout this book, the way things seem is often a poor guide to how they actually are".

His conclusion is that, since he cannot explain the interaction, the best way to solve the problem is to declare that the mind is a physical entity. To declare the interaction null and void. He says that *"conscious experiences <u>seem</u> non-physical"*. That is clearly not so. Conscious experiences *are* non-physical. I, at least, have never touched or smelled a conscious experience. Saying that conscious experiences have any of the qualities of physical substances is absolutely nonsensical. And cognition—a major part of human consciousness—does not originate in anything physical. We can say with all certainty that cognition is cultural. Maybe he only means that the mind is completely produced by physical processes in the brain. But that does not make sense either. How can that be when we have not created the words we use, nor do we experience sensations without often verbally sharing the sensation with other individuals, who understand exactly what we mean? How can it be when biology by itself cannot produce words? How can you touch or smell a conversation?

If Seth wants to say that consciousness is generated in the brain, that does not explain the nature of consciousness either. But language and body do intersect. Otherwise, how does Seth propose to classify the collective understanding of meaning?

LINGUISTS HAVE LONG KNOWN that a speech system has two different facets, one is abstract and the other one is physical: the abstract component includes semantics, (the different meanings of words or sounds), morphology (that is, how different words can be derived from a core of meaning, like "grace", and "gracious", and "graciously"), and the social context where sentences are uttered or written (which could be more respectful/formal or more informal depending on the circumstance).

But we also physically articulate sounds (like vowels and consonants, with many variations); and we vocalise making our vocal cords vibrate (or not) and using our lungs and throat (or not using them). In terms of discourse, we sometimes add intonation, rhythm and we can stress certain parts depending on areas we need to emphasise.

There is also another dichotomy. One component has to do with the emission and reception of sound and the other one with meaning. Interestingly, this understanding predates the formal field of linguistics and can be traced back to ancient Greek thinkers.

We are also aware that human phonetic ability has a lot to do with our articulatory apparatus and with the evolution of the elongated human larynx. No other primate can equal our command of communication. Our vocal production and the precision with which we emit vocal signals is incredibly engineered and precise. Again, something unequalled in the animal world. The u-shaped hyoid bone, for instance, which is located at the base of the oral cavity, is attached to the tongue and connects it with the epiglottis, the larynx and pharynx; that bone allows all kinds of movement and is involved in the productions of many specific sounds. It is not as developed in chimpanzees. Apparently, *Neanderthals* had a hyoid similar to that of modern humans, which made some scientists believe they could speak as well. Currently, there is debate about that.

A sound-emission system like that requires a reception structure to match it. But then, the human auditory equipment is up to the task. All this anatomical and neurophysiological refinement did not happen overnight. It is a product of the continued biological evolution that took place after humans underwent the great cultural leap.

It is evident, then, that our hominid ancestors did not have a linguistic equipment as advanced as ours. They were the ones who evolved it, through countless generations, from a much more rudimentary apparatus. However, once a hominin began to imbue sounds with semantic significance, a finite set of sounds naturally exploded into an infinite array of symbolic combinations. This intricate interplay of symbolic combinations—what we now refer to as language—likely predates, and is part of a feedback loop that aided in the expansion and emergence of our distinct cerebral cortex. In this sense, we might even posit that language not only gave rise to our advanced larynx and neocortex but also played a pivotal role in the emergence of our species.

Language is part of the intra-and-inter-cultural communicative process required to produce and maintain a universal human consciousness. Without language there would be no humanity. *H. sapiens*—the species named "the man who knows"—wouldn't know.

WE SAY that speech is produced by many organs: the larynx, the lungs, the mouth; that is the way language is externalised. But language is mostly processed in the neocortex, especially within the centre of Broca (located in the left hemisphere and used in speech-production and articulation), and the centre of Wernicke (in the temporal lobe, mostly to do with comprehension), which are connected through a neural pathway unique to our species and which we have known about since the nineteenth century. There is also the angular gyrus, which associates words we perceive with images, ideas and sensations.

The brain—then—is the central physical place where all these functions of language, communication, sentience and cognition integrate and, in term, produce behaviour.

When we discuss language, we should also include memory, which is facilitated by the hippocampus—that is, we can locate our cognitive and communicative abilities within many parts of the brain. That is where part of the process is located but, logically, it cannot be generated in the individual brain alone.

THE TITLE of a recent article by journalist Denise O'Leary in *Mind Matters*, an eMagazine, reads *"Does the brain constrain the mind instead of creating it?"* O'Leary concludes: *"There is no systematic, science-based reason today to think that's not a reasonable interpretation. And plenty of evidence suggests it."*

Well, very much like most of the things being said nowadays regarding the issue of mind and consciousness, the conclusion does not seem quite right because the question is incorrectly framed. The individual brain is one of the physical parts of a complex system that produces collective consciousness. Without the individual human brain there would be no individual mind, nor would there be a collective consciousness. Does a turbine constrain flight? Of course, a turbine is heavy, but without it there would be no propulsion, and therefore, no flight. Does a boarding pass constrain my ability to board a plane. Yes, it does to a point. Without it, it would be much easier to board a plane. But the boarding pass is part of a collectively agreed system that makes it easier to organise the flight and allow passengers to board in a certain fashion, and be assigned a given seat. O'Leary goes on to quote religious neuroscientist Michael Egnor. Egnor refers to the moment when the

INTERSECTION WITH BIOLOGY

soul is freed from the brain and from the body. That moment —he says—the individual has *"an enormous enlargement of experience, enlargement of knowledge"*. Well, whoever wishes to believe that the soul experiences an upsurge in knowledge the moment it is freed from the brain is entitled to their beliefs and opinions. There is not much more one can say about that.

I would say, though, that the brain does not generate consciousness, but neither does it constrain the mind. It is not a separate entity. The brain is an instrument of the mind. And the mind is part of human consciousness like a wave is part of the ocean.

The individual brain, as a component of the central nervous system, provides the individual all the sentience capacity the individual needs to survive, but is also one of the elements that create cognition and the language system we use to communicate with other human beings.

A description of how language is physically produced by the individual is really not necessary here. There are plenty of books and research on basic linguistics which describe those processes. What we need now is a deeper knowledge of the link between consciousness, language and culture, an area that has been neglected by science and that does not seem to be progressing because of the monistic fad among scientists that appears to have become the new dogma.

COMMUNICATION

"I seem to feel Napoleon's influence on our quiet evening in the garden for instance — I think I see for a moment how our minds are all threaded together — how any live mind today is of the very same stuff as Plato's & Euripides. It is only a continuation & development of the same thing. It is this common mind that binds the whole world together; & all the world is mind."

- Virginia Woolf, *Journal*, 1 July 1903.

𝒜s they speak, listen, read, or write, human individuals use words that are available to them from two sources, whatever other people around them are speaking or writing, and terms from their own lexicon. The latter—what linguists call the "idiolect" of that person— is not actually that private, because initially the mother, or parents, or the collective, taught that individual the terms he or she uses. When humans are born, they bring nothing to this world, not even words; any word they use, they had to learn. Throughout their lifetime, however, individuals acquire a thesaurus that depends on their level of intelligence, literacy or culture, and the people that converse with them. Very often, regardless of age, when individuals communicate, they learn new terms. The opposite side of the coin is that the language of a group of people is an aggregate of the many idiolects that compose that group.

From the above analysis of the linguistic interactions of an individual in a social group, it becomes evident that the consciousness of that individual takes into account two different cultural ingredients, a private one and a social one. That complicates even further the psycho-somatic layers we discussed in the introduction. A reductionist, biological, study of the brain cannot ever hope to capture that process in its totality.

Ortega y Gasset explains the complexity of serious communication:

"When the conversation is not a mere exchange of verbal mechanisms in which men behave almost like gramophones, when the interlocutors truly discuss an issue, a strange phenomenon occurs. As the conversation advances, the personality of each one of them progressively dissociates from it: part of it listens to what is being said and participates by saying something, while the other, attracted by the subject of the chat, like a bird by a snake, retreats into its own and thinks about it. When we have a conversation, we live in society, when we think, we are left alone. But the thing is that in that type of conversation we do both at the same time and, as the chat progresses, we do both at an increasing rate: we listen to what is being said... and at the same time we submerge ourselves into the deepest loneliness of our own meditation".

So, these are the two levels of communication we have been discussing; as we speak, we may be thinking about what is being said or we may be thinking about something else. We might even be trying to deceive our interlocutor by keeping quiet, or we might be outright lying to them. Exchanges of communication between humans reach levels of complexity that are impossible within other species. In biblical terms, we know good and evil. We have lost the innocence of the animal.

Human languages are immensely diverse and have many different types of sounds they use to communicate, i.e., clicks, and even whistles—like the clicks of the Xhosa, of South Africa, or the whistled languages of the Canary Islands, Greece or China, that are normally used to transmit information across valleys or long distances. The Hmong villagers of Long Lan, in Laos, a jungle people, transmit information using the sound of a special type of bagpipe, or sometimes even with leaves from plants they find in the forest.

The other layer of our communication system is that we have learnt that those sounds, that we emit orally and receive audibly, may also be represented by visual symbols. These come in different types: hieroglyphs—or ideograms—like the Maya and the Egyptians had, or the Chinese and Japanese languages currently use; syllabaries, like the hiragana and katakana systems in Japanese; and alphabetic symbols, for instance, the one we inherited from the Sumerians, through Greek and Latin. Ahh... and we also have Arabic numerals, which are mostly shared by different cultures to represent precise ideas of quantities. There are many more types of hieroglyphs that humans understand regardless of culture: nowadays we have emojis and icons, which are symbols that were introduced by electronic media in order to avoid extra typing through a single depiction of something or someone.

All of this is possible because we intuitively know that—even though any speech sounds like something fluid to the non-native speaker—that language is formed by discrete and combinable units of meaning, i.e., it is analysable and can be represented in many different ways. How did all that start? Well, the scratched image of an arrow on the bark of a tree, or on a rock, I imagine, could easily and intuitively be interpreted as signifying direction.

The caves of Europe have numerous examples of dots or bars—that seem to mean quantities—next to drawings of prey, like bison or deer. They represent seasons before parturition of the prey, when the animal is most vulnerable, a mnemonic system that would have aided the hunter.

What is evident is that there has been a beginning and a growth in our human consciousness. Alva Noë (*The Entanglement*), again, tells us that if we try to go too far back, to a time when our ancestors were more primitive, more basic animals, we will not find ourselves. In other words, that there is no point in looking for human consciousness, from an evolutionary perspective, before the appearance of human culture:

"We, that is, we psychologically modern H. sapiens, are the ones who talk, and cook and dress; we use tools and make pictures. It is here, amid this repertoire of skillful, technological organization, that the human mind, our distinct manner of being alive in and to the world, shows up. Go back too far, in the hopes of explaining who or what we are, and we lose ourselves."

The complexity of communication is a distinctly human trait. It has been culturally acquired; human consciousness—which includes our cognition—reflects that fact.

Attempting to study human consciousness using an evolutionary homogeneous framework ignores the rich historic and prehistoric multiplicity of our cultures.

As we have seen, the Gravettians, a hunter-gatherer pan-European culture were thought, for a long time, to be homogeneous throughout their territory. They were, actually, many different groups.

~

How do we begin to communicate with our parents and the collective? Babies and toddlers imitate behaviours and sounds. Children and teenagers keep on imitating, and then emulating, their parents and their peers. We even continue doing it as adults. We learn.

Our value system is the direct result of how our parents brought us up, and also of whom we chose to be our peers during our teenage years. If our parents do not teach us that stealing is bad, we may end up being thieves. If we see our parents drink to excess, then, maybe we'll be drunkards ourselves. Recruitment for the gangs of Central America, the 'maras', was conducted mostly among teenagers. It was 'cool' to do what your peers did. Role-modelling is one of the ways we learn to behave in society. A large part of who we are is cultural.

The biological elements of our consciousness—the inner components of the triune/adaptive brain—are the ones that come best equipped to make us begin to function when we are born. The neocortex, that uniquely human part of our brain—the component of the human brain that is roughly thirty times larger than in other mammals—is the portion that comes least equipped. The exterior of our brain adopts the behaviours that the collective teaches us.

At birth our bodies are ready to eat, defecate and urinate... and to cry. We are not equipped to do anything else. Giraffes can walk minutes after birth. Humans take approximately one year to learn to walk. We are an altricial species. The intake that comes from culture, i.e., behaviour, is extremely complicated and it takes years to learn. Of course, that includes language.

Mirror neurones, the neurones that drive us to imitate other humans, are present in different parts of the brain, especially in the neocortex. We constantly use those areas of the brain.

The results of the absence of human role-modelling can be quite amazing. Through history, there have been several cases of children who grew up surrounded by animals and learned their behaviours. One of the most recent ones is that of a Ukrainian woman, Oxana Malaya, who was born in Kherson Oblast, of alcoholic parents. When she was three, her parents left her outside to fend for herself. The toddler looked for warmth and protection among the dogs that were around the house. She was literally raised by those dogs. She lived in a kennel. All her contact with other beings occurred there. She became one of the pack. By the time a government agency rescued her, she no longer behaved like a human. She walked on all fours, and could not speak. She barked and growled and, in general, her behaviour was that of a dog. After rehabilitation, Oxana has a limited vocabulary and her mental capacity is that of a five or six-year-old. Unfortunately, she lost the opportunity to learn how to behave like a human being at the time when she required that role-modelling. An essential part of growing up is to be socialised when we are toddlers.

TIME AND CULTURE

"In the wordless beginning, space-time itself was squeezed and squeezed into a little ball in which everything gathered: what we call the singularity. Even if sound had existed then—it didn't exist, of course, because sound is made of matter—everything would have existed at the same time. Infinite amounts of every possible note would have been playing at the same time—the antithesis of music. Time was only born because that single point of totality stretched into a line and suddenly there was continuity. Suddenly, one moment began to distinguish itself from the other—the strange gift of entropy, which makes possible the existence of melody and rhythm, chords and harmonies."

- Maria Popova

What Maria Popova says is beautiful. I don't agree with her, but it takes the reader to an imaginary place where sound is made of matter and time is born. And it is fascinating. That is what many people want to believe. She says sound is made of matter, and I find that hard to believe. Sound waves are not matter. They may be microphysical, but they're not solid or tangible. They are immaterial. Maybe sound exists within a gooey present without which it would not be able to exist. Maybe sound is there because there is somebody who hears it, without whom it would not exist either. Maybe time was not born then. What probably began then, if ever, was change. I believe time appeared and evolved with human consciousness and it only exists within human consciousness.

Actually, the concept of time that I attempt to express in this book follows the Augustinian principle that *"time is an extension of the soul"*, which nowadays would translate as "time is subsumed in consciousness". Without our consciousness there would be no time.

Henri Bergson believed that time could be divided into lived time (real) and clock time (abstraction). From my perspective, time, as such, does not really exist. Lived time is—basically and fundamentally—change. Clock time is a creation, a way to measure the flow, the occurrence of, say, half a minute: what could be described as almost imperceptible change.

We've been measuring time since 'time immemorial' (pun intended). Perhaps I should explain what I mean: a recent article by Bennett Bacon *et al*, *An Upper Palaeolithic Proto-Writing System and Phenological Calendar*, published by Cambridge University Press, describes research on paintings, conducted in hundreds of European caves, and on engravings of bones.

The objects of the study were depictions of animals—prey—by *H. sapiens* some 37,000 years ago. For a long time, the depictions were believed to be art. It has become evident that they were not just art. They were mnemonic and notational devices. Maybe I would not call them proto-writing, but the study found that frequently occurring signs—like dots, lines and "*Ys*", paired with shapes of animals—were meant to carry meaning. The symbols signified months and seasons; they were part of a calendar beginning in spring and recording lunar months. The "*Y*" sign was indicative of parturition of the particular animal next to the notation.

What is fascinating about the research is that it demonstrates, among other things, that human beings, since that early stage, had been measuring lunar months—time—for hunting purposes. That is, that time (change and the repetitive nature of seasons) had preoccupied humans for strategic purposes since the beginnings of human consciousness. The finding convinces me even more that there is a close association between the emergence of human consciousness and the concept of time.

In the Chapter on 'Language, Culture and Consciousness', I discuss the fact that there are very few exact correspondences between languages. Here's a good example: in Spanish we have a very good word: "*siglo*". The English equivalent is called "century". But, originally, the two didn't mean exactly the same. Century is a word derived from one hundred—from Latin, "*centum*". That's a precise number, whereas "*siglo*" was not originally a precise term. And what does "*siglo*" mean? Where does the word come from? It comes from "*saeculum*"—diminutive of "*saecum*" (an era)—actually an Etruscan word that Romans adapted into Latin. It seems to mean "a short time, a few years," or perhaps "the memory of someone who is still alive": a lifetime, much better expressed in English as "within living memory". The Romans sometimes assigned it ninety years; others, up to one hundred and ten. Better to give an example to clarify what I mean. In his short story *The Witness*, Borges tells of an old man who is dying in a stable in England. The bells toll, calling for prayer. England is already Christian. But that old man remembers the god Woden, whom he had seen as a child; a crude wooden idol with Roman coins, to which horses, dogs and prisoners were sacrificed. When the old man dies, the last person who has seen the pagan god will die. When he dies, humanity will have lost something. Perhaps Borges did not remember—or was improbably unaware of—the hidden meaning of the term *"siglo"*, but what he says in the story is just what the Etruscans were trying to say. At that moment, the idol had become part of the distant past, it wasn't '*within living memory*' any longer.

Part of our collective consciousness is also a collective memory, because we think about the present, but we also know the past and think about it. We may not have historical certainty as to whether Jesus or Homer existed (and I have no bias either way). In any case, we know that Julius Caesar lived. Brutus and the other conspirators killed him. Let us forget what Shakespeare wrote: they were real people—like the one who is reading or like me—who saw it and who heard Mark Antony's eulogy after the magnicide. Tacitus and Plutarch wrote about him.

In Spanish we have adapted a new word like *"siglo"*, with its flexible meaning of time and we have made it a precise, scientific word, like "century" is in English. Now it means exactly one hundred years. That is how long it takes our planet to circle one hundred times around the Sun.

The most incredible thing is that nowadays many people are interested in finding out what "time" really is. After Einstein, the concept became "space-time", although we will see that his new idea, apart from adding an element, is much more complex than that.

Throughout history, and really because we needed to, we came up with increasingly sophisticated ways of measuring time. The

Egyptians and Sumerians had already divided the year into twelve months of thirty days. They also divided the day into twenty-four hours.

Of course, even if we assume that the way we measure time is accurate, that is also not true. Our concept of time is very limited and we have been inventing it little by little. As we have been inventing how to measure it.

For the ancient Hebrews, the day began when it was possible to distinguish a black thread from a white one. The days ranged from dawn to dusk. The idea had an indisputable logic to it. As it is indisputable that the Sun rises every day in the East and sets in the West: that happens three hundred and sixty-five times a year. That is, our planet rotates on itself those three hundred and sixty-five times as it circles the Sun. So far, so good. Our scientists have discovered that and we know it's a fact. A fact that we can even measure.

There are many other things that seem true but are totally arbitrary. The only certainty about them is that we are used to them.

A year has twelve months. A year has fifty-two weeks. This is the year 2024. A week has seven days. Really? Are those facts or total inventions? Well, the way we divide time is—by the way—a mere convention.

We have naturalised the divisions of time in such a way that it is already instinctive.

We say that planet Earth is 4.5 billion years old. And we measure the time of its existence in years because it's the only way we can begin to understand that amount of change, of movement. That is: a year is the time it takes the Earth to orbit the Sun. But 4.5 billion years ago—if there is anyone who can understand a figure of that magnitude—neither the Earth nor the Sun existed.

ARISTOTLE SAYS that time consists of two parts, since the present—the now—is not a part of it (I am not quite sure I understand this; perhaps what he meant was that the present was ever changing, i.e., something that becomes). He also says that time is not absolute but relative. The future is going to be at some point, but it still isn't, and the past is no longer there. Therefore, for him, the existence of time is somewhat doubtful.

Aristotle asserts that there is a relationship between time and change, but that they are not the same thing. Time is not change itself. After much rumination, he comes to the conclusion that time equals the amount of change. *"Time is the 'measure' of change"*. But he also explains that rather than a measure it is a type of order. He then comes to the following conclusion: time depends on the soul, but it can exist without the soul because it can be numbered and numbers are eternal. Maybe that last part should have been left out, as numbers only exist within human consciousness. Numbers are one more creation of the mind to be able to understand reality. There are no numbers in nature. There are different counting systems in different cultures.

Then, he also says that time can only be counted if there is someone to tell it. In the end, we agree: he says that without human consciousness there is no time.

According to physics, if something can be measured, quantified, defined mathematically, is an observable quantity on which other observable variables depend, that something exists. The question is: where does it exist? If I haven't made it clear thus far, the answer is that it exists only within human consciousness.

To be real, something has to meet all those conditions. In physics, if something is not possible, it is called a "pathology." Maybe time is pathological? Maybe it's impossible? Well, it meets all the conditions we listed before to be possible. I think it would have to be real. The problem is that the answers to all those questions are relative. They depend on where you are or if you are moving or if you are in the same place. That's what Einstein proved. Time is real, although it wouldn't seem to be real in an objective way. How?

According to the theory of relativity, time is not pathological. It is only relative. That means that if we put together the person who is on the train with the person who's still at the station, their times will be different, but if we do it again and again, with different people, the results will be consistent. The results are going to be predictable.

So, the idea that "time is relative" does not prove that time doesn't exist, or at least that's what it seems to me. In any case, what we can say is that time may exist, but it's just a construct; a measure we use to better understand the way some things work, and to function ourselves within society.

Carlo Rovelli, the eminent Italian physicist, writes in his book "*L'ordine del tempo*" that granularity is a universal characteristic: "*Perhaps the rivers of ink that have been spent talking about the nature of the 'continuous' through the centuries, from Aristotle to Heidegger, have been wasted. Continuity is just a mathematical technique to approximate things of very fine granularity. The world is subtly discrete, not continuous. The good Lord did not design the world with continuous lines: he did it with a light hand, he sketched it with dots, like a painting by Georges Seurat.*" Although we may not agree, it must be said that the man explains it in a brilliant way.

"Planck time" —the tiniest kind of time—has its equivalent in space: "Planck space", which is equivalent to 10-33 of a centimetre, one millionth of a trillion of a trillion of a trillion millimetres. Things like that are studied by quantum mechanics. Indeterminacy is another characteristic that is discovered at that level. You can't predict when—or whether— an electron is going to appear. Between one appearance and the other, the electron does not have a certain position. It disperses in a cloud of probability. This is called an "overlap." As quantum mechanics explains, space-time fluctuates and can overlap in different configurations. Again, Rovelli tells us: "*There is no single time. There is a different duration for each trajectory; and time passes at different rates depending on the place and speed. It is not directional: in the elementary equations of the world, the difference between past and future does not exist; its orientation is only a contingent aspect that appears when we look at things and do not observe the detail. In this blurred view, the universe's past is in a curiously "particular" state. The notion of 'present' does not work: in the vast universe there is nothing we can reasonably call 'present'.*" ... "*[Time] jumps, fluctuates, materializes only by interacting and, below a minimal scale cannot be found... So, after all this, what's left of time?*"

IN OUR LIVES, however, we need time. Our lives *are* constant change, and we need to measure it. We live now, learn from the past, and plan for the future. Are we the only species that does that? Absolutely.

Rovelli says there is nothing mysterious about the absence of time in quantum gravity. What happens is that at the fundamental level there is no variable for time. Time, at that level—as we have already seen, is what they call "Planck time"—so minuscule that it is negligible. Fantastic, that's as far as I understand.

But it also says something that I found very interesting. As an admission that time is not totally absent in quantum mechanics: when you change the position of a molecule, so does the state of the molecule. The same thing happens when changing the speed: there is another change in the state. But if you change the velocity first and the position later, the state of the molecule changes in a different way. That in quantum mechanics is called "noncommutativity" (I googled "incommutativity" but it doesn't seem to exist). In other words, the order of the changes cannot be altered without altering, also, the state of the molecule. Rovelli says *"La non-conmutatività determina un ordine, e quindi un germe di temporalità"*. At the quantum level there is a germ of time! It seems physics has a bit of poetry to it. But, is it true? Is that germ a germ of time?

Perhaps that is only putting a bit of feeling into something that is simply devoid of it, like physics. Of course, science can tell us, through its daughter, technology, how fast an airplane is going, its weight and mass, or why we can't have our cell phone on during the flight. What it can't tell us is whether it would be good for us to go see aunt Eugenia, who is very depressed. Nor can it tell us if we should yell at the neighbour to silence his dog. On the contrary, philosophy, art, poetry, music can give us some idea of what is right and what is wrong, or what makes us feel good or bad.

Perhaps there was a germ of time in the minds of the cave artists who wanted to leave a record of how prey behaved according to seasons for future hunters. One thing is for sure: there is no place in the animal brain where there are correlates for time. Since plants and animals appear to perceive only change (and perhaps, you could say, time to a very limited degree), since we have no biological connection to time other than cycles, the first thought that comes to my mind is that time is conscious, and also cultural.

WE HAVE SEEN—AND I have hinted at the beginning of this chapter—that time has preoccupied human beings throughout history, and even during some prehistory, as explained. But that preoccupation does not extend to other sentient animals. Why, if we share some form of consciousness with them?

The answer lies in the discreteness of the two consciousnesses. Our psyche appears to have introduced and gradually refined the notion of time. Again, why is that? Because human beings, in using our unlimited capacity for learning from experience, developed memory and imagination. We have long-term memory and we can apply strategy to our actions. Time is the tool we need to measure the long-term memory and the expectations we have developed. We need time to live in Western society, for instance.

As individuals, we learn about time from our parents, from the collective, from the culture, from the language we use. Time is important to us as a human collective. The way it develops in human minds is individual but it is derived from society. Time is taught again and again to every child and every generation. It does not come as an instinct. We learn it.

A baby is a human being, but in a stage of development. As such, a baby has very little autobiographical memory; he, or she, feels the passing of time like any other animal. The baby does not know, nor does she care, what happened last month. She cannot imagine the future at all. What's more, she cannot imagine, full stop. But when that baby becomes a little older, when she is a toddler, she begins to understand time. There is a today, a yesterday and a tomorrow.

IF TIME IS a construct of human consciousness, it follows that it does not disappear inside black holes. Time is not granular. Time may or may not bend in space, it depends on who's studying it, if anybody is. Physicists can speculate all they like, but time is only a human device, like numbers and identity. That seems quite evident.

Time—like dreams—is an ethereal creation of the mind. The difference is that dreams are individual and unconscious, whereas time is a conscious tool used by the collective to operate on a day-to-day basis, and to have a clearer idea of when memories occurred and when expectations will occur, if they become a reality. For all we know, our cave ancestors only thought in terms of seasons and the parturition of their prey. That was all they needed.

My assumption is that Alexander Selkirk—the real Robinson Crusoe—didn't need to measure time in order to survive on his island, except for knowing when days and nights and seasons would occur, and, if he did measure time further than that, it was just as a remnant of his life within human society.

The numbers we need to measure time exist only as far as we need them. Some cultures and some languages use numbers and time to a very limited extent.

In summary—I repeat—time exists; but it exists only within human consciousness. When there is no psyche, or when culture is in a very elementary stage, all there is is change. Animals know change, not time. Aristotle would have agreed with that. Human beings did not discover time. Human cultures invented it.

Neuroscientists have not discovered the correlates of time within the brain. Maybe they are there, but we use time because we need it; it is not innate, however. It is learnt. It appears obvious to us that time passes steadily, consistently and that we can measure its flow; and we know it goes from past to future.

WE ARE CREATURES OF HABIT. Habit and routine are papier-mâché versions of immortality we have created in order to survive without going crazy. We take time for granted; we are used to it; time has become ingrained in our minds. But time had a beginning. The beginning of time I refer to is _not_ the scientific one Thomas Hertog or Stephen Hawking meant when they wrote *On the Origin of Time* or *A Brief History of Time*. The one I refer to is the real one.

One of our ancestors—the last hominin that had not quite developed a human consciousness—had no idea of time. He most probably understood cycles, like nights and days, and seasons. But he could not think about a remote past. His memory (his hippocampus), was not as developed as ours. He could not think about the future. He had no imagination. The beginning of time did not happen during the Big Bang. It is a cultural and linguistic one.

Without time there was no history. We know that there was a prehistory, but we don't know what happened then. That knowledge has not been passed on to us. The task of archaeologists is to try and put pieces together to find out how things were.

Part of the evanescent nature of individual human consciousness is that, when the physical body disappears, so does consciousness. Unless something written or recorded stays behind, unless something we have created remains behind, we die and our identity, personality, our voice, our accent, our sense of humour, what makes us ourselves, all of that, disappears. Note that all of those traits I mention go beyond the individual, they are elements of the way we integrate in society. Even though these factors are highly individual, anything that makes us *us* has to do with society—something that Anil Seth (*Being You*) appears to reject.

What has come down to us before written history, the intermediate step, is just myth and/or legend. At one point, when there was no writing, there was oral history. We know that is the case because there are clear indications of how that happened.

Time is something we find very difficult to define, but something we measure. The way we perceive time is definitely individual and conscious, but mainly used within a cultural context.

In order to reinforce the main point of this chapter, let's revisit this: we acquire our notion of time through the conscious input of our culture. I believe understanding that becomes easier when we think about dreaming. Wakefulness and sleep are indeed part of our biological nature. Being awake and conscious we are aware of the passing of time. Asleep, we are not. When we dream, our minds, without time, unravel the cosmos. We know we are ourselves, we have self-awareness but —dreams being only subjective—we have no identity as such. Among other things, we may know some of the people in our dreams, but normally we cannot see their faces. Without logic, the time of our dreams does not appear to flow. It jumps, and turns, and returns, in unexplainable bursts.

Time in dreams seems to operate the same way Rovelli describes quantum time:

"[Time] jumps, fluctuates, materializes only by interacting and, below a minimal scale cannot be found... So, after all this, what's left of time?".

IN A 2021 ARTICLE titled *Time consciousness: the missing link in theories of consciousness*, two psychologists, Lachlan Kent and Marc Wittmann, argue that researchers do not give real 'time consciousness' the representation it deserves as an important element in human consciousness:

"Consciousness and the present moment both extend in time. Experience flows through a succession of moments and progresses from future predictions, to present experiences, to past memories.

However, a brief review finds that many dominant theories of consciousness only refer to brief, static, and discrete 'functional moments' of time.".

Current consciousness researchers depart from the premise that human consciousness is part of a continuum of evolution that goes back to other, very primitive, species. That is why researchers restrict themselves within certain limits. Human time, however—as opposed to animal 'time'—is almost unlimited for reasons other than evolution.

Kent and Wittmann appear to be on the right path:

"While there is a prevailing consensus in the field that consciousness is extended in time (Northoff and Lamme 2020), in our opinion as dedicated time researchers, it is not yet extended enough. Decades of timing research supports a 'minimally sufficient' duration for time consciousness somewhere in the seconds' range (Fraisse 1984; Pöppel 1989, 1997; Varela 1999; Wittmann 2011; Kent 2019), but most theories and methodologies in consciousness science only focus on the hundreds-of-milliseconds' range (Northoff and Lamme 2020)... One possible misconception at the root of this problem is that time consciousness is synonymous with the timing of behavior, perception, and other stimulus-based responses or event-based experiences.". It would seem that current neuroscience is only concerned with 'animal time'.

Kent and Wittmann propose:

"Given the lack of work dedicated to time consciousness, its study could test novel predictions of rival theories of consciousness. It may be that different theories of consciousness are compatible/complementary if the different aspects of time are taken into account. Or, if it turns out that no existing theory can fully

accommodate time consciousness, then perhaps it has something to add.".

Adding long-term or autobiographical memory to their research would probably force neuroscientists to realise that human consciousness is the product of a big leap, and not part of an evolutionary continuum. That, in turn, may dissuade them away from pure physicalism and into more interdisciplinary work.

TRANSLATION

*A*ll translators know that one-to-one correspondence does not exist in any pair of languages. Because of that, translation is not an exact science. It is closer to an art. Not only that, languages are alive and move like—and with—human groups. Habit and recursiveness displace languages ever so slightly over the centuries until mutual intelligibility disappears. This is reflected in space and time. Cultures change, and with them, their languages tend to vary. Romance languages are Latin no more. Politics, geographic accidents, and history have operated changes in them and divided them into mutually unintelligible entities.

Sanskrit is not Proto-Indo-European, both of them, to a larger or lesser degree, are lost in the mist of time. Four thousand years ago, Sanskrit became diluted into her daughter dialects—the Prakrits—spoken in the northern areas of India, and layered over Dravidian languages now totally unknown. European languages and cultures—which share the same common Indo-European ancestor—could not be more distant from their

Indian relatives. Greek, and Etruscan, and Latin among other older European languages, were battered by waves and waves of Eastern invasions that changed both the languages and the mentality of their speakers. Southern European languages varied even further by adopting massive amounts of lexicon from Arabic. Maltese is an extreme example of that phenomenon.

NINETEENTH CENTURY SWISS linguist Ferdinand de Saussure described the differences between *langue* and *parole*, the former being the abstract structure—the actual operating system—of a language, and the latter, the way the language is uttered, the speech that corresponds to that language. As a translator, I have no doubt that both, the rules of a particular language, and the utterances produced by any speaker of that language, are cultural and—in turn—create the linguistic loop that will influence that culture and, through it, its individuals.

Poetry—largely untranslatable—is probably the essence of a written language. There are great examples of poetry in translation, though. In his book "Cathay", which he produced with help from linguist Ernest Fenollosa's notes, taken from Chinese poems, Ezra Pound created exquisite jewels (*"The river merchant's wife: a letter"*; *"The lament of the frontier guard"* to name two); he also wrote poetry and Noh plays taken from Japanese. Pound was fluent in French and Italian and could dabble in the lyrics of those languages in a more direct fashion. Without taking any merit from Pound's poems translated from Chinese and Japanese—and accepting that the sound and the music could no longer be there—one wonders how faithful they are in terms of nuance. Without arguing against their beauty, one wonders whether the concepts could be reflected in English (and understood by English speakers), so alien are the culture and the language.

Could a *gaijin*, a foreigner, write good *haiku* in proper Japanese? I suppose it depends. Only if the *gaijin* can think like a Japanese native speaker, I imagine.

Noh theatre, would be a much bigger ask. It would be even more difficult than a Japanese speaker trying to understand the Nordic sagas.

Having said that, translations are possible. The only thing that is questionable is how much the translation retains from what the original means to a native of that culture.

Having explained—say, to a Tok Pisin speaker from Papua New Guinea—who Jesus was, and how Christianity spread throughout Europe, how much of a Bible that speaker would understand? It would probably take a lifetime to explain the origins and different meanings of the Old and New Testaments.

Do we think in different ways depending on our culture? Do we have a different worldview? Is language a force that shapes our thoughts? Trying to prove that scientifically would take rigorous empirical testing.

Would proving that Linguistic Relativity is correct imply anything in terms of our common humanity? Would it have any racist implications or consequences? I very much doubt it. However, any attempts at proving the hypothesis would be faced with a lot of opposition from that front (universalists clamouring against racism, etc.).

At least from the times of Neolithic Europe, humans appear to have had cultural differences. Ever since the early beginnings of humanity, speakers of different languages showed different regional and cultural mentalities. Were the differences unbridgeable? It does not appear to be so, as there are human beings now in the billions and we all communicate. There is no doubt that what unites us is the humanity we all share. In any case, all humans can reflect. We may have different perspectives, but we enjoy the use of our cognitive and metacognitive abilities, which other animals do not appear to have.

WHEN I WAS STUDYING linguistics at the Australian National University in the 1970s, all the rage was (the then-almost-unfalsifiable) universality of languages. I remember the insistence on the universality of semantic primes, for instance. And having to do two compulsory undergraduate courses on generative grammar. The implication was that formulas could be applied to language and that any sentence could be generated using those formulas. The principle didn't appear a sound one to me. After I graduated, I discussed my doubts with Professor Hammarström, at Monash University, in Melbourne, and discovered that I was not the only one that believed that Noam Chomsky had sold a dud to linguists. I was naturally inclined towards a Sapir-Whorf, cultural kind of thought, and could not find how the intricacies, nuances of a language—the idiosyncracies of a culture—could be explained by reducing it to formulas. I still believe that language is cultural and that both, language and culture, largely influence the cognitive functions of human consciousness and—in a strange feedback loop—are reflected by them.

Having said that, one of the most distinguished lecturers at the ANU School of Linguistics at the time, now an Emeritus Professor, Anna Wierzbicka—a semanticist—was the originator of the Natural Semantic Metalanguage (NSM) theory. Her approach maintains that there is a small group of semantic primes that can explicate the meaning of culture-specific notions and values. The idea becomes an extremely useful tool for cross-cultural communication in the hands of linguists. MSN has all kinds of specialist applications and does not contradict in any way the views expressed here in terms of the cultural vs universal theories concerning consciousness.

A RECENT STUDY from Tokyo University of Science used colexification (a method that analyses indirect semantic associativity) to establish that four central concepts that are emotion-related: "good", "want", "bad", and "love", are universal. According to the study, those words have the highest amount of correlations with other emotional terms in many languages. The researchers identified emotional hubs that exist across languages. The study, qualified as a "breakthrough" in the article, intends to demonstrate that sentience is universal to all human beings. That is not news. Of course qualia terms, and anything to do with emotion, are universal. Languages are still cultural: knowledge, information, cognition-related terms are part of the input a human receives from culture as a child.

Using the same colexification method, another study reported in a 2019 article—this time from the University of North Carolina at Chapel Hill—compiled a database of 100,000 terms from over two thousand languages from twenty major linguistic families. Many of the words, emotional and nonemotional, were discovered to be polysemic. The semantic fields of words vary from language to language, culture to culture, even from variant to variant, or among dialects of a language.

According to Angeles Carreres, a translation expert from Cambridge University:

"Much of this variability stems from languages' history and cultural context, which are rarely conveyed by cut-and-dried translation dictionaries."

Unfortunately, stating that language and culture may influence thought is often interpreted through the discourse prevalent in nineteenth-century scientific circles. In those days, scientists assumed humanity followed a hierarchy in which Western people were often situated at the top. Different peoples had reached different stages of evolution and that was blamed on their cognitive abilities. In summary, human beings were not inherently equal; and colonised peoples were inferior.

But language and culture influence cognition, and nineteenth-century ideas on the hierarchy of races do not necessarily follow.

NEED FOR A DEFINITION

"...Every ego so far from being a unity is in the highest degree a manifold world, a constellated heaven, a chaos of forms, of states and stages, of inheritances and potentialities. It appears to be a necessity as imperative as eating and breathing for everyone to be forced to regard this chaos as a unity and to speak of his ego as though it were a one-fold and clearly detached and fixed phenomenon. Even the best of us shares this delusion."

—— Hermann Hesse, *Steppenwolf*

To begin with, the title of this chapter requires further exploration and explanation. When we discuss a basic *"dictionary"* definition of consciousness, the task of producing one sounds relatively simple. In fact, there are already many English language 'definitions of consciousness'. When I say 'English-language definition', I do it on the basis that much of the research has been already conducted in English. What is meant here is that there has to be a universally-agreed English-language definition. Not only that, there should be several definitions of consciousness-related terms and phrases, and they should have clear limits. That poses serious scientific and philosophical problems and would require a concerted effort of great magnitude.

Terms and phrases like mind; subconscious; understanding; experience; unconscious; conscious; brain; sense; sentience; comprehension; speech; intelligence; cognition; metacognition; nervous system, to name a few, would require a more detailed, more precise definition.

The fact that "consciousness" has not been properly defined means that it often is equated with "wakefulness", as in being "conscious" or "unconscious", when that should only mean being "awake" or "asleep". Actually, the main difference between wakefulness and sleep—and there are other, important ones—is that, when the individual is asleep, there cannot be communication with other individuals. There are different levels of sleep; we are not going to enter into their description, though. Neither are we going to discuss the effects of anaesthetics and the deep sleep they cause.

But to begin to define something we need to understand what we really mean by *"defining"*. The etymologies of *"defining"* and *"definition"* take us to Latin through Old French: the term *"dēfīniō"* means "to establish limits". To start with we need to find limits, to define and classify each one of them; it is not going to be an easy task. The referent, "human consciousness", as we have seen through the pages of this book is a process, a function. One of the problems that have been found in the study of consciousness is that the complexity of the subject covers many disciplines and affects the way in which, historically, Western science has studied things. Not only is the process physical (and intangible as well), but it has sub-functions. The sub-functions are also hybrid in terms of their physicality, but also in terms of their objective-subjective nature.

∼

APPARENTLY, the current discussion of whether LLMs *"truly understand the world"* or do not understand it has resulted in a possible solution to the problem. Amazon Web Services (AWS) have advanced some definitions that are uncontroversial and provide a consensus concerning the terms used in that particular subject. According to Stefano Soatto, of AWS, *"These definitions are useful in the sense that they align with definitions that people have used in mathematical logic, as well as mathematical model theory, as well as epistemology, dynamical systems, and linguistics and so on and so forth".*

In a recent article, Trager and Soatto argue that it is possible for LLMs to 'understand' meaning:

"Suddenly, mainstream media are alive with speculation about whether models trained only to predict the next word in a sequence can truly understand the world. Skepticism naturally arises. How can a machine that generates language in such a mechanical way grasp words' meanings? Simply processing text, however fluently, would not seem to imply any sort of deeper understanding."

But they continue: *"... For a given sentence, we consider the probability distribution over all possible sequences of tokens that can follow it, and the set of all such distributions defines a representational space. To the extent that two sentences have similar continuation probabilities—or <u>trajectories</u>— they're closer together in the representational space; ... Sentences that produce the same distribution of continuations are "equivalent", and together, they define* *an <u>equivalence class</u>. A sentence's meaning representation is then the equivalence class that it belongs to."*. That applies to sentences, but a similar method can be used for words.

So, if a conclusive, logical, definition of 'meaning' can be reached using something called the 'distributional hypothesis' (the distribution of words is closely related to meaning) and computers can evaluate those distributions, then, the meaning of any term can be assigned an 'objective' value, or something as close as possible to it.

A GOOD EXAMPLE of why we need clear definitions is a recent study conducted by Julia Notar, a doctoral candidate at Duke University. Notar trained brittle stars to anticipate food. Brittle stars are brainless creatures with a basic nervous system. What Notar did was similar to what Pavlov had done with dogs: she habituated the stars to eat when their environment was dark. Once darkness came, the brittle stars would come out even before any food was in their tanks. They had linked darkness with food. Notar's conclusion was that her brittle stars had undergone a process of associative learning, and that therefore they enjoyed cognition.

The problem with Notar's conclusion is that any basic definition of cognition requires a brain. To mention one definition: *"Cognition is a term referring to the <u>mental</u> processes involved in gaining knowledge and <u>comprehension</u>."*. The operative words here being "mental" and "comprehension". The stars did not have a brain, neither did they understand that there was a problem to be solved. They became habituated, pretty much like Pavlov's dogs, but without a brain. Habituation can happen with even more basic forms of life. To some extent—

and depending on the definition of the term—we can safely state that high cognition is a human phenomenon.

Perhaps the best places to start researching definitions of consciousness and consciousness-related terms would be graduate schools of computer coding, linguistics, and philosophy. What's more, semantics perhaps should include a unit dedicated to the origin of human consciousness, maybe with input from neuroscience and paediatric physiology. The acquisition of language by toddlers is one of the areas related to consciousness which has been largely ignored by neuroscience.

To give an example, sentience—one of the two major divisions of human consciousness—is basically subjective in that it is very difficult to ascertain, at an individual level, if the colour you see is the same as the colour I see, or if the pain I feel is the same as your pain. I say "at an individual level", but we also find that the meaning of certain subjects has changed historically and culturally: we don't know if Homer saw that "wine-coloured sea", if he was referring to a blue wine, or if ancient Greeks, as a collective, could not perceive certain hues or certain colours. There are, of course, cases such as Korean speakers, who find that perceiving differences between certain shades of blue and green is difficult, and many other examples of different nuances affected by culture, but we are not going to enter into that now.

We do have basic standards and we can communicate what we mean in terms of sentience although, in general, a sense will remain basically something that the individual feels, be it emotions, or input through body organs, like the nervous system or the brain.

CONCLUSION

Perhaps we should try to recapitulate, even if it is partially, in order to refresh concepts and make some sense of what we have said.

At the end of the *Introduction*, the book summarises the most important, points of the hypothesis: that human consciousness consists of two integrated but discrete layers, one of them biological (sentience), with which we are born; and the other one, cultural (psyche), which each generation of human individuals acquire through contact and care from relatives and the collective.

The cultural component of the hypothesis tends to coincide with—and corroborate—linguistic relativity: language and cognition are transmitted culturally which, to some extent, would tend towards falsifying any universalist claim. The next step points out the futility of any attempt at understanding human consciousness using a totally evolutionist, reductionist, framework, as it would find the ceiling presented by culture. Evolutionist research will necessarily have to stop when it

CONCLUSION

reaches the most recent point in the biological evolution of the individual and the beginning of culture in the earlier, previous species: the point where one species transitions into the other, i.e. the birth of *H. sapiens*.

I then list all the individual human traits that are acquired through contact with the culture. There are quite a few, and they make us different from any other animal species. Two of them are long-term memory and imagination; those two help us open up the gates of past and future, which are closed to all other animals. I conclude that time is a human construct that we invented to meet the needs of civilisation, and that without human psyche there would be only present and change.

In the following chapter, I attempt to summarise what happened at the 26th meeting of the Association for the Scientific Study of Consciousness, in June 2023. Maybe what I describe is not as clear as I expected it to be. What I try to explain is a feeling I have that, to an outsider like me, the whole episode appears to have been somewhat confusing.

The neuroscientific bias against dualism has created an incredible atmosphere in science. Some, like Kristof Koch and Giulio Tononi, were proposing the 'scientific' equivalent of panpsychism as an answer to the question of human consciousness. All theories appear to have been acceptable to the scientists. Some of these notions, like 'a conscious universe', were totally unfalsifiable. Their current approach, then, falls short of providing explanations for the concept of a collective consciousness, and neuroscientists have shown limited interest in exploring cultural and sociocultural phenomena.

In certain instances, the insights from quantum mechanics, particularly those related to wave function and entanglement, appear to have been misinterpreted and uncritically embraced as corroborating evidence for a monistic paradigm. Notably, within the realms of neuroscience, philosophy, and physics, some theories—like Tononi's IIT—have propounded the idea that consciousness is an intrinsic and universal element of matter, and that it is measurable. This perspective has swung the pendulum heavily in one direction, sometimes leading to strong reactions against alternative viewpoints.

It is worth acknowledging that there is broad recognition of the interconnectedness of systems in the universe, but adopting monism in its strictest sense—asserting that all matter in the universe possesses consciousness to explain phenomena like entanglement in physics, or the inability to pinpoint the emergence of consciousness within the confines of the brain—seems to defy logical reasoning. Some call it pseudoscience. Karl Popper would be turning in his grave.

In the chapter about the philosophical problem, I try to describe the nature of human consciousness; I question the fact that all research being conducted by neuroscience is based on the individual brain. Western science appears not to acknowledge the existence of the collective mind, of the culture. Neuroscientific research has, thus far, concentrated its efforts on the biological side of consciousness: sentience.

I believe the cognitive and metacognitive components of human consciousness should be studied as a partially collective phenomenon. The individual transmits that phenomenon through biological means. But it cannot be confused with an individual phenomenon. Only its expression is individual. Human individuals and their behaviour cannot be studied in isolation.

The chapter then turns to Large Language Models, AI, and systems like ChatGPT, and explains how these are shallow

CONCLUSION

expressions of ideas that are input into them by different means. The machine cannot be conscious because it is not alive, it cannot understand the culture that created it. It only adopts what has been implanted into it without any real comprehension. Any AI system must be superficial because it is only a secondary iteration of linguistic input. Natural language transmits ideas. It does not perceive them, and it only shares part of the sentient component of consciousness. The AI output appears to show that the automaton understands and feels but, lacking the live, biological layer, to some extent it parrots very complex ideas and expresses them, often impeccably. Eventually, though, if not real consciousness, it may reach some understanding.

From LLMs, which basically are a semblance of cognition without biology, the chapter then veers towards the early life of the human individual: a live being that has not as yet acquired cognition.

The chapter closes with the thought that Western science needs to consider the possible existence of the meta-evolutionary, layered, dual, hybrid, nature of human consciousness from a social perspective before any further advances can be made.

The following chapter, *Language, Culture and Consciousness*, which is actually the central chapter of this book, touches on the different levels and functions of consciousness and how the communication process operates. It includes a short discussion on linguistic relativity. And it does so, because one of the premises of my hypothesis is that *psyche* is culturally transmitted, which tends to support the principles of linguistic relativity.

After a lifetime devoted to the study of languages and translation work, there is no doubt in my mind that culture and language are inextricably intertwined, and that the result of that intangible admixture is reflected in individual human consciousness, (and in the synapses of its physical component, the brain, of course). Cultural relativity seems obvious to me (if politically incorrect nowadays): cultures do solve problems in different ways.

In the chapter I have just mentioned I explain that—although translation does indeed occur—oftentimes many of the cultural nuances are lost to the interlocutor at the receiving end of the translation or interpreting process.

I also mention a study conducted by researchers in Spain and the United States, based on the Young Finns Study, which appears to corroborate the hypothesis of this book: that human consciousness consists of two discrete but intertwined layers (or networks of genes expressed in the brain); one of them is biological (*"anxiety, fear, etc."*) and the other one is cultural (*"e.g., production of concepts and language"*).

To summarise what I have expounded so far in this book, I believe that—as opposed to what happens with any other animal species—one of the main components of consciousness in human beings is cognition. If we accept that individual cognition is the result of a meta-evolutionary process which began with and includes sophisticated culture and recursive language (both of which definitely lack in other species), the logical conclusion is studying human consciousness acknowledging its synergetic nature, which—as opposed to animal consciousness (basically sentience)—would not emerge within the individual brain, but rather it would emerge in the collective. Thus far, science has successfully used reductionist methods, but they are not to be confused with something fundamental. They cannot account for the synergy that comes from the collective, especially in language and culture. Once you start analysing the nature of a rope and need to cut it into segments sufficiently small for the purposes of the study, the 'ropeness' of the rope is lost. In this case, it is not just the reductionism that becomes destructive: the most important component of human experience is being ignored. To be successful, then, any new study of human consciousness would need to start afresh, and be conducted with a proportionately large input from the humanities.

When I say that culture and language are lacking in other animal species, I recognise that other living beings may have some kind of communication and that some species may even have community-based behaviours, like chimpanzees, elephants and some birds. What I mean, then, is that *H. sapiens* remains the only living species to have developed a large, sophisticated cultural structure, based on the use of natural languages with very broad lexicons and recursive syntax.

Any new definition of human consciousness would have to include linguistic and cultural components in order to allow a more comprehensive study of the subject.

I realise that what is being proposed here would run afoul of the principles of normal Western science, which has been traditionally physical and based on the individual. Western science and philosophy share indeed a common ground: "objective reality".

CONCLUSION

However, science has so far failed to elucidate any of the subjects that are part of the "hard problem" of human consciousness; maybe the reason for that is "objective reality". But maybe it is because the discipline in which the study of consciousness has been concentrated is mostly neuroscience, whereas the venues of research could perhaps include schools of linguistics, mainly areas like semantics, but also philosophy, psychology, sociology, and anthropology; in any case: the humanities.

The main reason human beings have survived and thrived as a species is the fact that we have proved to be much more gregarious than any other species, *H. Neanderthaliensis* included. We are—I repeat—a highly altricial species. We need long periods of care and preparation that enable us to behave within society.

We are like flocks of birds and schools of fish. Individuals are very important to our species, especially within the context of Western culture. Because of that we need to understand that the main component of consciousness is the one that is intangible and shared.

Beginning the study of human consciousness from a completely different perspective would take a lot of new analyses. What appears to be an insoluble problem that has taken decades to resolve merits to be considered from a much more inclusive sociocultural, humanistic perspective.

And we have to remember that language would be a crucial element in any future study of human consciousness because it necessitates physical and non-physical elements from the individual. Language is dual and hybrid in that it includes both

CONCLUSION

biological and social agency. All of it is important, from neurones and synapses, to points of articulation, to meaning and comprehension.

Reactions to the reductionist approach traditionally adopted by science—which seems to have run its course—are beginning to pop up: Adam Frank, Marcelo Gleiser and Evan Thompson have recently published *The Blind Spot: Why Science Cannot Ignore Human Experience* (2024). If what is needed is an explanation for human consciousness, Western science—based on "objective reality" as it is—cannot provide any of the answers.

Frank *et al* describe the problem in a recent article published online in the Big Think: the study of human consciousness marks the limit of objective reality.

"In biology, the origin and nature of life and sentience remain a mystery despite marvelous advances in genetics, molecular evolution and developmental biology... Cognitive neuroscience drives the point home by indicating that we cannot fully fathom consciousness without experiencing it from within. Each of these fields ultimately runs aground on its own paradoxes of inner versus outer, and observer versus observed, that collectively turn on the conundrum of how to understand awareness and subjectivity in a Universe that was supposed to be fully describable in objective scientific terms without reference to the mind."

In their book *The Blind Spot*, they conclude:

"In short, although we have created the most powerful and successful form of objective knowledge of all time, we lack a comparable understanding of ourselves as knowers. We have the best maps we've ever made, but we've forgotten to take account of

the map makers. Unless we change how we navigate, we're bound to head deeper into peril and confusion.".

Even though Frank, Gleiser & Thompson's main reaction to the current scientific approach has to do mostly with reductionism and objective reality, I identify wholeheartedly with their conclusion. I would add the discrete and integrated layers that allow your brain to interact with mine. I would add that you—the reader—and I are part of a unique meta-evolutionary phenomenon.

I commenced this book by quoting a Zen poem.

> *"On a withered branch,*
> *a crow has come to perch—*
> *at dusk in autumn."*

I believe I should mention here, perhaps, for those who are not aware of it, that buddhist philosophy rejects cognition. Therefore, it cannot be explained intellectually, it can only be explained through absolute experience. It preaches that the only reality—which buddhists call '*nothingness*'—is just perceived through sentience: i.e., reality as it was before humans became 'dual'. That is what Bashō attempts with his beautiful description. He reproduces an experience.

When a Zen monk reaches '*satori*', he becomes part of '*real*' reality. He reclaims his part within that reality; he '*is*' the garden, the flower, the arrow; he does not witness from the outside any longer. To be enlightened, then, (to reach the ultimate wisdom), the monk learns to reject cognition.

CONCLUSION

In doing so, however, in separating those two components, Zen philosophy appears to corroborate the hypothesis of this book. Human consciousness includes sentience *and* cognition.

Whether buddhist monks decide to live only through sentience; whether they believe that perceiving '*real*' reality and reaching wisdom is only achieved through sentience, that is another matter.

ACKNOWLEDGMENTS

Inés, as she has done before on several occasions, assisted me at all stages of this project. She knows I am most grateful for that. This book is dedicated to her.

I thank Mark Lush for providing help with much appreciated information (a great book), for reading the draft, editing, and for his comments.

Rod Haedo offered his observations and suggested amendments on several occasions. Most of them are included here.

Mike Hetherington undertook to read a second manuscript and came up with very good ideas. Much appreciated.

Kirsty Gillespie read an earlier draft and helped with useful suggestions and contacts, which I mention in the book. I am grateful for them.

My brother Patocho commented and helped on several occasions.

Lachlan Kent was a dissenting voice among those who read the manuscript. The book was toned down after receiving his criticism. I have already expressed my gratitude to him.

SOME QUOTATIONS

"...Why would it matter, he asked, if machines someday surpassed humans in intelligence, even consciousness? It would simply be the next stage of evolution. Human consciousness, Musk retorted, was a precious flicker of light in the universe, and we should not let it be extinguished. Page considered that sentimental nonsense. If consciousness could be replicated in a machine, why would that no be just as valuable? Perhaps we might even be able someday to upload our own consciousness into a machine. He accused Musk of being a 'specist', someone who was biased in favor of their own species. 'Well, yes, I am pro-human', Musk responded. 'I f***ng like humanity, dude'."

—— Walter Isaacson, *Elon Musk*, 2023.

∽

"It seems plain and self-evident, yet it needs to be said: the isolated knowledge obtained by a group of specialists in a narrow field has in itself no value whatsoever, but only in its synthesis with all the

rest of knowledge and only inasmuch as it really contributes in this synthesis toward answering the demand, 'Who are we?'"

—— Erwin Schrödinger

～

"Our universities fail to guide us down the easiest paths to wisdom... Rather than teaching a sense of awe, they teach the very opposite: counting and measuring over delight, sobriety over enchantment, a rigid hold on scattered individual parts over an affinity for the unified and the whole. These are not schools of wisdom, after all, but schools of knowledge, though they take for granted that which they cannot teach — the capacity for experience, the capacity for being moved, the Goethean sense of wonderment."

—— Herman Hesse

～

The traits I respect the most are erudition and the courage to stand up when half-men are afraid for their reputation. Any idiot can be intelligent."

—— Nassim Nicholas Taleb

～

"I am not sure that I exist, actually. I am all the writers that I have read, all the people that I have met, all the women that I have loved; all the cities that I have visited, all my ancestors."

—— Jorge Luis Borges, *El País interview*, 1981

SOME QUOTATIONS

"Our scientific worldview has gotten stuck in an impossible contradiction, making our present crisis fundamentally a crisis of meaning. On the one hand, science appears to make human life seem ultimately insignificant. The grand narratives of cosmology and evolution present us as a tiny contingent accident in a vast indifferent Universe. On the other hand, science repeatedly shows us that our human situation is inescapable when we search for objective truth because we cannot step outside our human form and attain a God's-eye view of reality."

—— Adam Frank *et al*, *The Blind Spot*, 2024

"Whatever understanding is, it must be beyond computable physics... it's not necessarily beyond science. It's just beyond current science. My claim is much worse, much more serious, much more outrageous than that it's quantum mechanics in the brain. People say 'quantum mechanics in the brain, oh, it can't be that'. That's worse. I'm saying, that's where quantum mechanics goes wrong. It's a theory which we don't know yet."

—— Sir Roger Penrose, *New Scientist interview*, 2024

"If someone says it's raining and another says it's dry, it's not your job to quote them both. Your job is to look out of the f***ing window and find out which is true".

—— Jonathan Foster, Professor of Journalism.

www.ingramcontent.com/pod-product-compliance
Lightning Source LLC
Chambersburg PA
CBHW051444290426
44109CB00016B/1671